Naturalistic Environments in Captivity for Animal Behavior Research

SUNY Series in Endangered Species
Edward F. Gibbons, Jr. and Jack Demarest, Editors

Naturalistic Environments in Captivity for Animal Behavior Research

Edward F. Gibbons, Jr.,
Everett J. Wyers,
Everett Waters, and
Emil W. Menzel, Jr.,
Editors

STATE UNIVERSITY OF NEW YORK PRESS

Cover photo: White-handed gibbon in the Jungle World rainforest habitat at New York Zoological Park. Photo by E. Gibbons.

Published by
State University of New York Press, Albany

© 1994 State University of New York

For information, address State University of New York
Press, State University Plaza, Albany, N.Y., 12246

Production by E. Moore
Marketing by Theresa A. Swierzowski

Library of Congress Cataloging-in-Publication Data

Naturalistic environments in captivity for animal behavior research /
    [edited by] Edward F. Gibbons, Jr. . . . [et al.].
        p.   cm. — (SUNY series in endangered species)
    Includes bibliographical references and index.
    ISBN 0-7914-1647-X (alk. paper). — ISBN 0-7914-1648-8 (alk.
paper)
        1. Captive wild animals—Housing—Congresses. 2. Captive
    mammals—Housing—Congresses. 3. Primates—Housing—
    Congresses. 4. Animal behavior—Research—Congresses.
    5. Primates—Behavior—Research—Congresses. I. Gibbons,
    Edward F. II. Series.
SF408.45.N38   1993
636.08'31—dc20   CIP   92-39108
                                                              CIP

Contents

PART V. CONCLUSION

Preface

This volume is the result of a 1987 conference: "Non-standard Indoor Environments for Naturalistic Behavioral Research," which was cosponsored by the National Science Foundation and the State University of New York at Stony Brook. Since that conference, the book has evolved to consider both indoor and outdoor naturalistic environments for animals in captivity. The basic premise of the 1987 conference, however, remains the same: to foster a greater understanding of the roles and responsibilities that administrators, animal behaviorists, curators, federal regulatory personnel, and veterinarians have in the design, construction, and operation of naturalistic facilities for animals in captivity. The present volume also emphasizes the scientific need for naturalistic environments, and the value such habitats have for species conservation and public education.

The book is divided into four sections. The Introduction section outlines the purpose for the volume, and highlights many of the theoretical and pragmatic considerations involving research in naturalistic captive environments. The second section, Animal Care and Use: Regulations and Case History, outlines current federal standards for the operation and inspection of naturalistic facilities, and reviews a case history concerning research in these nonstandard habitats. The third section, Physical Dimensions and Management Considerations in the Design of Naturalistic Environments, considers the impact of physical environmental parameters on the health and well-being of animals, examines strategies for the control of arthropod pests, and discusses management considerations in the design, construction, and real-time operation of these facilities. The fourth section, The Importance of Naturalistic Environments: Research on Selected Species, focuses on scientific findings from species of birds, rodents, and primates. These investigations emphasize the value naturalistic habitats can have for basic and applied animal behavior research.

Continuity between the four sections of the book is provided by the general theme that behavioral research in naturalistic environments can substantially promote the health and well-being of animals in captivity. Specifically, behavioral research in naturalistic environments can enhance our understanding of the biological and psychological functioning of animals by addressing questions difficult or impossible to examine in the wild, including issues regarding species conservation. The volume, therefore, challenges the notion held by some animal activists that no research should be conducted on animals in captivity. The book also challenges research institutions to better provide for the biological and psychological needs of animals. It is evident from the discussion presented in each of the volume's chapters that valid behavioral research can best be achieved by using animal subjects whose physiologic and psychological functioning arc within the limits defined as normal for the species. Naturalistic environments in captivity can be important methodological instruments in the achievement of valid behavioral research.

VOLUME EDITORS

Acknowledgments

This volume is the culmination of efforts on the part of individuals associated with the 1987 NSF/Stony Brook conference, and the present book. The conference organizers and editors of this volume extend their appreciation to the faculty and staff of the psychology department at the State University of New York at Stony Brook, in particular: Drs. K. Daniel O'Leary and Edward Katkin, past chairs of the department, Dr. John Neal, Ruth Shepard and Beverly Rivera—past and present assistants to the chair, respectively. The editors would also like to thank the secretarial staff of the department: Carol Carlson, Debbie Campani, Barbara Conklin-Lento, Paulette Gerber, and Ginny Munnich. A note of appreciation is also extended to Judy Smith of Travel Vision who coordinated the travel arrangements for all conference attendees, and to Richard Kachigian and Donna Schoenfeld for their able assistance during the conference. Everett Wyers, Everett Waters, and Emil Menzel would like to thank their respective wives—Lee, Harriett, and Harriett—for their support during the organization of the conference.

The present volume could not have been possible without the dedication of the editorial staff at the State University of New York Press. In particular, a special note of appreciation is extended to Priscilla Ross for her attention to detail and encouragement throughout the preparation of this volume. The editors would also like to thank colleagues who took time from their busy schedules to provide peer review of the volume's chapters: Drs. Robert Cancro, S. Sue Carter, J. Derrell Clark, Susan Clarke, Barbara Durrant, John Fleagle, Nelson Gernetta, Michael Hutchins, David Johnson, Carl J. Jones, Devra Kleiman, Fred Koontz, Robert Lickliter, Martin Morin, Melinda Novak, R.S. Patterson, Michael Power, Edward O. Price, Michael Robinson, Benjamin Sachs, James Taylor, John Vandenberg, and Frans de Waal.

The editors extend a note of acknowledgment to Mr. James Doherty, general curator, and to the mammal department of the New York Zoological Park for hosting the first day of the 1987 conference. The tours and discussions of the naturalistic facilities at the zoological park contributed significantly to the scientific content of the conference and the present volume.

Behzad Barzideh, microcomputer systems specialist, and Nancy Duffrin, coordinator of instructional computing, at the State University of New York at Stony Brook are gratefully acknowledged for their assistance in preparing the book for publication.

Finally, the conference and preparation of this book was supported largely through a grant from the National Science Foundation (BNS 8610271).

The logo for the SUNY Series in Endangered Species was artfully drawn by Stephen Nash of Conservational International.

I. INTRODUCTION

1

Naturalistic Facilities for Animal Behavior Research: In Search of Accommodations

The monitoring and enforcement of animal care standards has increased significantly in recent years in response to advances in laboratory animal science (e.g., Driscoll, 1989; Mann et al., 1991; Orlans et al., 1987), increased awareness of animals' biological and psychological needs (e.g., Besch, 1990; Fox 1986; Moberg, 1985a; Novak and Suomi, 1988), ethical issues related to the use of animals in research (e.g., Remfry, 1987), and increased public awareness of abusive practices in animal research (e.g., Phillips and Sechzer, 1989). As a result, it has become increasingly important for researchers to understand the roles and responsibilities of all parties who become involved in the design, construction, certification, and operation of animal facilities. Naturalistic facilities, in particular, are facing difficult periods of adjustment to the current regulatory environment. In many cases research in these facilities will be temporarily suspended for redesign, changes in animal care and use practices, and expenditure of funds not associated with specific scientific studies. Equally significant, it is becoming increasingly difficult for researchers to establish new naturalistic facilities because of their nonstandard design, and the attention they draw to animal research in captivity.

The central premise of this volume is that understanding the dimensions along which naturalistic animal facilities diverge from regulatory standards, and understanding the perspectives and concerns of responsible parties at the institutional and federal levels can help researchers to more effectively advocate and conduct animal behavior research (Richmond, 1991). Such an understanding will also help researchers to articulate their needs while being cognizant of the complications these requirements may present for others in authority. These topics have administrative, economic, interprofessional, and legal facets to which animal behavior researchers are not often attuned, especially early in their careers.

The collection of papers in this volume address theoretical and pragmatic issues concerning naturalistic environments that will help researchers to think more strategically about nonstandard features they may want to build into these facilities. The contributions provide insights into identifying physical environmental features that may not be in compliance with existing regulations, and may have a negative impact on the physical health and psychological well-being of the animals (see chapters by Besch and Killias, and Stoskopf and Gibbons, this volume). The early identification of such physical environmental features will enable researchers to seek cost-effective remedies in the design plan or to obtain waivers for the operation of facilities that are at variance with the published standards. Finally, this volume will encourage students interested in animal behavior research to establish working relationships with veterinary and regulatory personnel, and to acquire basic sophistication in laboratory animal science issues during their graduate and postdoctoral training.

THE NEED FOR NATURALISTIC HABITATS IN CAPTIVITY

Traditional animal research enviroments were designed to optimize the control of and access to the animal subjects. An objective of these facilities was to minimize the time and financial cost needed to conduct research, to maintain the facility, and to care for the animals. The control and standardization of the animals' physical and social environments was considered essential to the attainment of husbandry and research objectives.

The price tag for this kind of access and control, however, is too high if it is purchased at the expense of attention to the psychological well-being of the animals. Accordingly, an important consideration in the design, construction, and operation of modern naturalistic environments is the range and quality of the interactions between the physical environment and the animals' physiological, morphological, and behavioral functioning. Such a design orientation has cast naturalistic environments in pivotal methodological roles in contemporary animal behavior research. For example, Collier and Rovee-Collier (1981), Kummer (1982), Menzel and Juno (1985), Shettleworth (1984), Shettleworth et al. (1990), and Yoerg (1991) have employed naturalistic habitats in influential studies of animal cognition. Research on social biology and behavior related to reproductive success have also emphasized the methodological importance of naturalistic facilities (e.g., Beck and Power, 1988; Frederickson and Sackett, 1984; Holmes and Sherman, 1982; Melnick and Kidd, 1983; Reiter et al., 1981; Spencer-Booth, 1970; Widowski et al., 1990). The range of adaptive behavior under study in naturalistic facilities is rapidly expanding.

The importance of naturalistic facilities is also being pushed to the forefront by recent efforts in several problem areas of applied animal behavior research. Some of the most pressing of these is related to the preservation of endangered species, which are increasingly having to be propagated in captivity (e.g., Gibbons et al., 1994). Compared to the advances in modern fertility research, efforts to reintroduce captive-bred animals into the wild have met with very limited success. One of the primary reasons for this is undoubtedly related to the inadequacy of standard captive-rearing environments for the development of normal foraging, social, and predator avoidance skills (Gittleman and Conover, 1994; Hutchins et al., 1994; Kleiman, 1989; Konstant and Mittermeier, 1982). Recent advances in interspecies embryo transfer techniques also pose challenges, which, as applied to endangered species, can only be addressed in naturalistic environments (Durrant and Benirschke, 1981; Gibbons and Durrant, 1987). In sum, naturalistic facilities will play an important role in research on the effects that rearing conditions have on subsequent behavioral adaptation. In addition, they may come to play essential roles as transitional habitats for diagnostic, remedial, and research purposes prior to the release of animals into the wild (see chapter by Beck and Castro, this volume), and in the advancement of behavioral medicine for both animals and humans (e.g., Gibbons et al., 1992).

ADMINISTRATIVE AND REGULATORY CONSIDERATIONS

In view of the importance of naturalistic environments in the advancement of theoretical and applied animal behavior research, it is unfortunate that rapid changes in regulatory standards, financial support for research, and social currents are making it difficult to establish and maintain these valued facilities. On the regulatory side, it is clear that published standards are intended to accommodate research requirements where this does not compromise the research animals' needs and security. At the same time, many more animals are used in industrial and biomedical contexts than in behavioral research, and standards are naturally addressed specifically to the types of housing employed in these settings. Conceptualization of nonstandard facilities and procedures for waivers and certification have received little attention, particularly in the published standards themselves.

Because violation of animal care regulations can result in serious repercussions within an institution, administrators are not eager to encourage development or expansion of naturalistic facilities, except in the few instances where behavioral research more than "pays its own way." More importantly, naturalistic facilities tend to require significant amounts of space, are expensive to operate, and often present security problems not encountered in conventional facilities. Thus, in order to generate administrative support for naturalistic research in captivity, researchers need to understand the concerns and responsibilities related to these demands. Further, it is critical that researchers consider how the methodologies proposed for research may conflict with administrative and regulatory policies, and to seek solutions that will yield valid scientific information and will not infringe on the physical health and psychological well-being of the animals.

Each year any number of individual papers on specific aspects of animal housing appears at professional meetings and in journals devoted to laboratory animal science and animal behavior research (e.g., Coe, 1989; Nicholson and McGlone, 1991; Novak and Bayne, 1991; O'Neill and Price, 1991). Many of these contributions address the limitations of current standards, the scientific rationale for nonstandard housing in specific applications, or novel solutions to housing problems. As more papers appear on the effects that environmental variables such as lighting, sound, temperature, space, or substrate have on animal behavior and health, one can easily determine that housing standards for a given species are inherently complex to write. Any useful set of standards will be controversial and in some cases restrictive relative to the researchers' needs. What is not easily gleaned from these papers is an overall framework within which to comprehend the situation as it applies to naturalistic facilities in captivity. Moreover, these technical papers generally do not address the practical and political issues that make it difficult to establish and operate naturalistic facilities in the current regulatory environment.

CONCLUSION

The following set of interdisciplinary papers will promote the identification and conceptualization of major technical, regulatory, and practical difficulties associated with the design, construction, and real-time operation of naturalistic animal facilities in captivity. Further, they will help administrators, regulatory personnel, and scientists to address the question: What constitutes a naturalistic environment for animals (see, for example, chapters by Dewsbury, Glickman and Caldwell; and Novak et al., this volume)? Discussion on this and similar topics is presented in the concluding chapter of the volume. This chapter is a product of a synthesis statement developed by the participants of the 1987 conference. The synthesis statement also considers the operational philosophies and educational benefits of naturalistic habitats, and provides a protocol for the resolution of conflicts regarding the regulation of naturalistic environments in captivity.

It is also intended that the following contributions will enable investigators to address their research needs to the full array of responsible parties in a timely manner. The papers will also help scientists interested in animal behavior processes to determine the most strategic avenues along which to design and advocate research requiring naturalistic facilities. The contributions might also warn off scientists who would undertake such research in complete innocence of the administrative, regulatory, and technical obstacles. It is also hoped that the volume will encourage students in animal behavior research to develop working relationships with veterinary and regulatory personnel, and to acquire basic sophistication in matters related to laboratory animal medicine and husbandry during their graduate and postdoctoral training. This experiencc is increasingly necessary if young investigators are to retain the initiative in designing their research, obtaining external support, and providing the best care for their animals.

ACKNOWLEDGMENTS

Work on this paper was supported in part by a grant (BNS 8610271) from the National Science Foundation. The volume editors also appreciate the hospitality and cooperation extended by the New York Zoological Society, in particular Mr. James Doherty, general curator, and the staff of the mammal department.

K. DANIEL O'LEARY

2

Whither the Naturalistic Study of Animals in Captivity? A View from the Chair

A cover story in the science section of the *New York Times*, "Dolphin Courtship: Brutal, Cunning, and Complex" (Angie, N., February 18, 1992), captured my interest. Drs. Smolker, Richards, and Conner have been studying the herding behavior of 300 dolphins over the last decade in Shark Bay, Western Australia. For 25 months they observed the behavior of dolphins from a 12 foot dingy, noting individual characteristics of dolphins by identifying scar patterns on their fins and by recording the sounds of various members of the group. As a clinical researcher interested in physical aggression in intimate human relationships (O'Leary, 1988), I found the report that male dolphins engaged in intricate group alliances to herd females of special interest. In contrast to humans, these male and female dolphins differed little in size, and several males were required to "capture" a female for mating. Once "captured," the female was subjected to controlling and aggressive behaviors by male dolphins such as charging her, biting her, swimming around her in unison, and making "popping" noises at her. It is not yet known whether a female dolphin mated with one or several of the males that ganged up to "herd" her because dolphin copulation was not observed by the researchers. The determinants of the aggressive or controlling behaviors of the male dolphins are unclear, but it is obvious that males engaged in very aggressive behaviors toward the female once she was captured. The researchers believe that a large amount of the variance in aggressive behavior of these dolphins resulted from the fact that the female dolphin has only one calf every four to five years and the fertile female is a very prized commodity.

Why is this research of interest? It is the study of a mammal known for its intelligence and social behavior. It is the study of social behavior of animals in a natural habitat. It is this type of research that makes me wonder whether my colleagues who wish to study animals in seminatural habitats created at university centers will be able to do so without unusual hassles and threats to their research.

One Saturday morning in the fall of 1985, Emil Menzel came to talk to me about problems he was having with the animal inspection teams who give approval to the habitats of animals at research centers. Dr. Menzel was conducting research on the social and foraging behavior of marmoset monkeys, and he used a university greenhouse for this research. The greenhouse was converted to a large open area in which the monkeys could roam freely and could climb small trees growing in the dirt floor. A veterinarian, Dr. Middleton, was the director of animal care for the university, and he had indicated great concern that the facility would not meet standards for animal care established by the U.S. Department of Agriculture (USDA). More specifically, he was concerned about the dirt floor, insects, live trees, and birds that flew freely in the enclosed greenhouse facility. The story that unfolded about the care of monkeys in the psychology department that were in the trust of Dr. Emil Menzel was ultimately the prompt for this book and the National Science Foundation conference that followed the closing of the university greenhouse facility for animal research. As such, the events surrounding "the Stony Brook Greenhouse Incident" deserve special attention.

THE STONY BROOK GREENHOUSE INCIDENT

Background Information

I was a past chair of psychology and was serving as chair of a dean's administrative council who administered the psychology department while we searched for a new chair. As you will see later, I think my role as a transition administrator was an unfortunate one, for I was not going to inherit the long-term problems of animal care, and I thus was not able to negotiate long-term plans for animal care facilities. Nonetheless, I was thrust into the middle of the problems of animal care in seminatural environments.

My early experience with animals was during high school on a Holstein dairy farm and in showing cows in Chester County, Pennsylvania, at the county fair and at the state farm show in Harrisburg. Later, as a graduate student at the University of Illinois and young faculty member at Stony Brook, I engaged in collaborative research in two conditioned suppression studies with rats as subjects (Van Houten, Weiss, and O'Leary, 1970). Having considerable experience with farm and laboratory animals did not at that time seem very relevant for coping with the problems of caring for monkeys in seminatural habitats. The issue raised by Dr. Menzel's research involving the housing of animals in a seminatural environment seemed easy to grasp. Housing monkeys in a large greenhouse with space for free roaming and social behavior seemed far better than having monkeys in large cages—even if those cages were easier to clean and were more sterile. It seemed to me that even a casual observer would judge that monkeys in a large greenhouse with the ability to roam freely and climb real trees were in a better environment than in large cages.

A number of parties were involved in making decisions about the care of the animals living in the university greenhouse. Initially, the key players were the following: (1) Dr. Emil Menzel and his graduate students, (2) the Department of Laboratory and Animal Care represented by Dr. Middleton, (3) vice provost for research, Dr. David Glass, a social psychologist and member of the psychology department, and (4) president of the State University of New York at Stony Brook, Dr. John Marburger. At the time of questioning the animal care in the university greenhouse, there was considerable unrest across the country about animals used in experimentation. Indeed, personnel at an animal facility at the University of

Connecticut, which provided for the naturalistic study of wolves, were harassed and the wolf facility was eventually shut down as a result of the activities of the animal rights groups (Personal communication, B. Ginsburg, University of Conn., December 15, 1992). Care at certain animal laboratories at the University of Pennsylvania had been found questionable and National Institutes of Health (NIH) funding to all laboratories at the University of Pennsylvania was threatened, according to the director of our animal care facilities, Dr. Middleton. In brief, there was a mood of tremendous unrest caused in large part by the threats of animal rights activists, and Stony Brook was attempting to get their "animal house" in order. One illustration of the unrest about animal care was the U. S. Senate bill (John Melcher, Montana) that was accepted on October 1985. Among other things, this bill sought environments that had exercise opportunities for dogs and promoted the psychological well-being of primates. In addition, to help guard the confidentiality of research, heavy fines were to be imposed on animal welfare reviewers who leaked information (Marshall, 1985).

As chair of the department's administrative council, I met with Drs. Menzel and Middleton to ascertain if there were ways to meet the department of agriculture standards for New York state as well as the federal guidelines regarding animal care. In the past, the greenhouse had passed inspection many times, though there had been some citations for violations that were corrected. Further, in 1985, a new inspector was designated to assess Stony Brook's facilities, and the facility failed inspection on September 19, 1985. After the failure of the inspection, our department discussed how to have the facility remain a naturalistic facility and yet pass inspection. Ordinarily there is a period after a failed inspection for the university to make changes and to have a reinspection. Around this time, the university switched to centralized animal care, which was a change from a period in which psychology had its own animal care personnel. These personnel had excellent familiarity with the purposes of the naturalistic research with the marmosets and tamarins in the psychology facility. Thus, the failed inspection may have been due in part to the new inspector, in part to the absence of an animal care worker in the department per se to communicate the purposes of the research, and in part to the zeitgeist of the times—namely, have every animal care facility meet every standard. In part, according to some people with whom

I spoke, it may also have been due to the lack of clear lines of communication and responsibility.

Closing of the Greenhouse

During ongoing negotiations with members of the administration and the head of laboratory animal care, Dr. Menzel and his graduate students were told that they would be forced to shut down in October 1985, and that the animals would be moved to caged facilities. The moving of these animals would essentially ruin the ongoing research of Dr. Menzel and his graduate students. Movement to caged quarters would require total acclimation to such new quarters and new learning in the vastly changed environment.

The official reasons for closing the animal care facility in the university greenhouse were listed as follows in a written communiqué to Dr. Emil Menzel: (1) problems of keeping a secure animal facility, (2) the presence of roaches, and (3) an issue of whether the facility was scientifically productive. I did not see the particular memo that communicated these points, but various memos had communicated concerns about these problems. The issue of security was noted because monkeys had broken free on at least two occasions in 1984 or thereabouts. However, it was clear that this issue could be solved by various fencing material around the outside of the greenhouse as a secondary barrier. There were definitely roaches in the greenhouse facility which Dr. Menzel later found emanated from the adjacent building. In my opinion, this problem could also have been surmounted. The productivity of Dr. Menzel's research group was a concern not in an absolute sense but relative to the cost of housing the monkeys. In 1985, the cost of caring for a marmoset monkey was 91 cents per day; given the 14 marmoset monkeys, the total cost per year to the department was $4,650. At the time, Dr. Menzel was a coinvestigator of a grant from NICHD to Emory and Georgia State Universities, though that grant did not support the greenhouse facility. In 1985, Emil Menzel reported to me and the dean's administrative council that he had provided significant personal money for equipment and his own personal publication costs. Nonetheless, the costs of his animal care were a significant part of the department's support costs and a cost that was high relative to that incurred by other faculty members. Only Dr. Logue, who had 24 pigeons, had a higher capital cost; the cost of animal care for a pigeon was 65 cents per day, and for the 24 pigeons, the annual cost was $5,694. While these costs per faculty member were considerable, when a faculty member applied for federal grants, as had Dr. Logue, we tried to support the faculty member's research.

In my opinion, there was a confluence of forces at work that led to the closing of the greenhouse facility. Most critical was the desire of the university to avoid any bad press and a possible loss of grant funds. In a letter to me of October 22, 1985, from vice-provost Dr. David Glass, he stated:

> When we know of problem areas that can be corrected to prevent jeopardizing our total teaching and research program, steps must be taken to accomplish this task even though it may impact on someone's research program. Such is the case with the marmoset colony. President Marburger has directed me to keep the greenhouse closed for the time being. . . . As for the marmosets, there are other facilities that are suitable for housing these animals. This space may not be as desirable from a research standpoint. However, it is imperative that we do all we can to comply with rules, regulations, etc. so as not to be cited by N.Y. and the UDSA inspectors, not violate our NIH assurance statement, nor our AAALAC accreditation.

The absence of any clear guidelines for animal care in semi-naturalistic settings enabled the university to act swiftly to close the animal facility in question. This closing was also made easier by the changing leadership in the psychology department, and probably by my lack of knowledge about the options that existed for the inspection and regulation of animal care facilities in naturalistic environments. In the absence of any clear guidelines with which I was familiar, I felt that steady negotiations were the best course of action. However, the university administration closed the animal facility shortly after the animal inspection failed.

RESPONSES TO THE CLOSING OF THE GREENHOUSE.

On October 28, 1985, Menzel wrote to vice-provost Glass that an official at NIH indicated that NIH had never cut all funds to investigators based on the actions of one laboratory. He also made clear that the central reason for the failure of the USDA inspection was that the university viewed his laboratory "as an ordinary animal colony area rather than as a special research environment for the naturalistic study of animal behavior and failed to draw to the attention of the USDA inspector or his superiors to Section 2.100 of the Animal Welfare Act. . . . How can one possibly study the naturalistic behavior of social groups of an almost completely arboreal, and highly insectivorous, species of monkey without having conditions that are 'nonstandard' by USDA criteria?"

On November 4, 1985, Dr. Menzel officially protested the closing of the animal facility in a letter to vice-provost Glass, Dr. Middleton, director of animal care for the university, and to me. The central point of his memo was that he was not given due process. He forcefully stated that the university had more than the two options that were stated to him by the university administration—that is, to "correct all existing deficiencies by October 4, 1985, or to move the animals out of Central Hall [the building to which the greenhouse was attached] and close the facility." He continued:

> Rather these were the only two options that the University itself chose to consider, and others did and still do in fact exist, as anyone sufficiently familiar with the Animal Welfare Act itself would surely know. They include a letter of intent to renovate, accompanied by a request for extending the deadline; requesting a variance to certain USDA specifications on the basis of the specific research aims of the studies in progress; and filing an appeal and a request for a reinspection in which the nature of the research and the particular species of research animal is more fully taken into account.

Finally, reflecting his frustration and anger, Menzel concluded: "NIH officials themselves would be far more apt to say that nothing in their decision-making process is automatic or cast in concrete, for their purpose is to guide and assist investigators in doing better and more humane research, not to put them all out of business."

Attached to the letter from Menzel were four letters sent directly to president Marburger in support of Dr. Menzel's research program and requesting a reconsideration of the university's decision to close the nonhuman primate facility. From what I can determine, these researchers heard the depiction of events from

Emil Menzel, and, as far as I know, their letters were based on what most would consider partial evidence. Nonetheless, these letters convey a sense of the great concern by highly regarded researchers about the closing of a facility for the seminaturalistic study of animals. Those letters were written by Dr. Stephen Suomi, chief, laboratory of comparative ethology, NIHCD, Bethesda, Md.; Dr. Charles Snowden, professor of psychology and zoology, editor, *Animal Behavior*, chair, conservation committee, American Society of Primatologists; Dr. Jeffrey French, departments of psychology and biology, University of Nebraska at Omaha; Dr. Suzanne Chevalier Skolnikoff, conservation committee, International Primatological Society, San Francisco.

In addition to protesting the closing of the greenhouse facility, these letters mentioned Menzel's international reputation for his recognition of the special cognitive abilities and special living requirements of nonhuman primates. Suomi stated: "His [Menzel's] greenhouse based concept for housing New World primates has been copied and successfully utilized in several laboratories in North America, South America, and Europe, such that in every case in which parallel data were available the greenhouse environments have been found to be superior to traditional laboratory environments conforming to AAALAC standards for mammals in general from the standpoint of general health and reproductive success."

When vice-provost Glass informed me in writing that the greenhouse facility was to be closed, I talked to him about the problems that the closing created for one or two graduate students who were then completing master's and dissertation research with the marmoset monkeys. I requested that these students receive extended funding for their graduate research and, if necessary, money to travel to other primate centers to conduct observations of monkeys in naturalistic settings. Dr. Glass assured me that funding for these purposes would be granted. In addition, the university administration sought outside consultation from Dr. Ben Beck from the National Zoo in Washington, D. C., regarding how to rehabilitate the greenhouse facility if the animal facility were to be reopened.

Following the closing of the facility, letters by many animal researchers to President Marburger at Stony Brook led the university to back down from its original position, and Dr. Menzel was to have the greenhouse facility rehabilitated to meet animal care standards. In turn, Dr. Menzel was asked by the new chair of psychology, Dr. Edward Katkin, to write a grant for his research, but by that time, Menzel was frustrated by the local situation. He decided to abandon his animal research on campus and instead spend more time in collaboration with colleagues at the Yerkes Primate Center in Atlanta.

INTERPRETATIONS OF THE MEANING OF THE GREENHOUSE CLOSING

Basically, Emil Menzel and his graduate students who studied the behavior of marmoset monkeys in a seminatural environment were caught in the crossfire of state and federal regulations, on the one hand, and pressure by colleagues not to give in to what they perceived as unfair treatment, on the other hand. In attending the meetings on animal care, it became clear that the primary regulatory codes were written for animals in closed quarters such as cages. Requirements such as cage sizes, number of animals per cage, ventilation, and the scrubbing apparatus for the cages seemed prominent in the regulations. Little was written about the care of animals in seminatural environments. Instead, as Menzel pointed out when care deviated from the norm, a special justification had to be made for the deviations. In hindsight, the university veterinarian and the university administration acted in a fashion that now appears to have been unnecessary from an animal care standpoint.

It seemed to me that the judgment of administrators was partly based on productivity of the greenhouse facility in terms of grant funding and published materials relative to the cost of maintaining and renovating the facility. However, if that were the case, I was not officially informed of those factors at the time of the closing of the facility. As noted earlier, however, Dr. Menzel told me that he received a written communiqué that questioned the productivity of the facility. The fact that a grant proposal was requested by Dr. Edward Katkin as a condition for the renovation of the greenhouse facility after it had been closed directly indicates that more effort in this regard was being requested or required of Menzel to seriously consider the reopening of the facility. At the bottom line, it seemed that a cost/benefit analysis was an issue that led to the closing of the greenhouse facility. A full discussion of that issue is beyond the scope of this chapter. In fact, in 1992, Dr. Glass, the former vice-provost for research, indicated to me

that the cost/benefit ratio was indeed a factor that figured promi-
nently in the decision to close the greenhouse facility. Administrators need to be aware of the significant costs of maintaining animals in seminatural environments. In many cases the decision to
support such facilities may be determined by the likelihood of federal support after initial university support has been granted. Such
an issue clearly had some influence over the administration in this
case.

FUTURE DIRECTIONS IN THE NATURALISTIC STUDY OF ANIMALS IN CAPTIVITY

In my opinion, from an administrative standpoint and from a
cost/benefit analysis, team research or research involving several
colleagues who use the same animals may be necessary to sustain
animal research in seminatural habitats. Given budgetary constraints, departments and universities will have to make decisions
about whether support costs for animal care, clinical facilities, and
experimental laboratories fit into the overall graduate training
needs of departments or interdepartmental efforts, especially if
these facilities do not generate private, federal, or state money to
sustain a significant percentage of the training and research needs.
Finally, state universities will also have to determine whether different university centers will have particular foci such as animal
research in natural environments whereas other centers may have
no research in that area. As I see it, for research purposes, small
graduate research programs with faculties of three or four profes-
sors in a number of campuses would often be better melded into
one major center at a single university site. (Such consolidation
would of course take time and consideration of faculty rights and
needs.) Even for undergraduate training, breadth may need to be
sacrificed for depth and an esprit de corps among faculty members,
which in turn may be communicated to undergraduates. In summary, both to encourage high quality research and faculty morale
and at the same time to be responsive to fiscal issues, state university systems should consider concentrating programs involving
studies of animals in natural environments in few or single locations.

The proposal to the National Science Foundation for a Stony
Brook conference, *"Non-standard Indoor Environments for Naturalistic Behavioral Research,"* seemed especially important to me

for continuing the discussion of this issue at the national level. The issues are ones that relate directly to the kind of research that one can conduct, the amount of training that is necessary for pre-doctoral graduate students who wish to conduct research in semi-natural environments, and the political issues that ensue if these problems are not well-thought out. Most importantly, the dialogue is necessary if guidelines are to be established when the existing guidelines for laboratory research do not apply. My sincere hope is that this book, which emanated from Stony Brook's NSF-sponsored conference, will prompt federal and state regulators to work with researchers who wish to conduct naturalistic research to develop general guidelines for the care of animals in such facilities.

The role of the greenhouse facility and Stony Brook's involvement in the study of animals in natural habitats was updated by Emil Menzel who gave me some letters regarding the closing of the greenhouse facility. I had requested certain letters from him for my introductory chapter for this book, *Naturalistic Environments in Captivity for Animal Behavior Research*. He noted that Rachel Smolker, who conducted the dolphin research I cited above, was the daughter of the late Robert Smolker, a former member of the SUSB department of ecology and evolution. She sat in one of Dr. Menzel's courses, "dabbled in the greenhouse," and still visits Emil Menzel when she comes to Stony Brook—further documenting Menzel's influence on students. Her advisor in graduate school was Dr. Richard Wrangham, a primatologist who wrote a letter of support to Dr. Marburger on Emil Menzel's behalf.

Having reviewed the correspondence about the care of animals in seminatural habitats and the strife that was caused by the problems attendant to this issue, I can only hope that chairs of psychology and related disciplines across the country will become aware of the options that exist for animals that are not housed in standard caged facilities. I also hope that this book will provide administrators of animal care facilities with the necessary understanding of the recommended guidelines that exist for animal care in both indoor and outdoor naturalistic environments for animals in captivity. I will do my small part by personally circulating this introductory chapter to all the chairs of psychology in the United States in hopes of alerting them to the issue and to the book that can provide some answers to the care of animals in seminatural environments. The book covers issues from control of anthropod pests to the significance of naturalistic environments for primate behavioral research. The concluding section provides guides for what are appropriate naturalistic environments in captivity for animal behavior research. The material in this concluding section should be known to all directors of animal care as well as to researchers interested in the study of animals in seminatural habitats.

ACKNOWLEDGMENTS

Thanks go to Drs. Edward Gibbons, David Glass, and Emil Menzel for their comments on earlier drafts of this chapter. People have different memories of the events surrounding the critical incidents depicted in this chapter, and I take full responsibility for attempting to integrate them. The opinions expressed herein are mine unless otherwise noted.

3

Comments on Behavioral Research in Naturalistic Settings

We can conceive of things only as they appear to us and never as they are in themselves: how things are in themselves remains forever and entirely out of the reach of our thought.
                                        Thomas Nagel, 1986

The welfare of animals must depend on an understanding *of animals, and one does not come by this understanding intuitively: it must be learned.*
                                        Sir Peter Medawar, 1972

These epigraphs pose a paradox that all who seek to understand animals and their behavior must face. On the one hand, science can never rest assured of the psychological well-being of any animal—or of any person—in any setting. On the other hand, the welfare of animals depends on knowing what constitutes animal well-being. In recent years, this dilemma has come into much sharper focus.

The heightened emphasis on animal well-being is based on the parallel development of two trends in human socialization which are as old as civilization itself. Progress in public concern for human and animal welfare accelerated during the past 200 years. So, too, did scientific study of animal behavior progress. As a result, these two factors have together come into clear focus (Fox, 1986). However, that focus has given rise to discrepant positions regarding animal welfare (Dawkins, 1983, 1984). It seems to me that the underlying theme of this book is that behavioral research in naturalistic settings promises to resolve this discrepancy by bringing those two conflicting positions together.

The discrepancy itself appears to rest in the limitations of science in creating consistent interpretations of acknowledged facts—interpretations often referred to as theories or models. The dilemma, then, is the need to bring those limitations into accord with public views of animal well-being. Thus, it may be said that the purpose of this volume is to examine the usefulness of naturalistic settings in furthering our understanding of animals.

Two questions are involved. Do naturalistic research settings facilitate development of useful generalizations about animal behavior and its causation? Is the animal better off—more representative of a healthy member of its species—in a naturalistic setting than in typical laboratory caging? What constitutes a proper answer to these questions is still another question. The answer may very well rest in the paradox posed by the epigrams that open this chapter.

This particular chapter is divided into two parts. The first offers a few thoughts on animal welfare and how well-being is to be judged. The second presents some considerations which limit my view of what comprises an ideal naturalistic research setting.

KNOWING VERSUS UNDERSTANDING

In our dealings with animals, we often fail to take note of what the epigraphs quoted in this chapter suggest. To know an animal is not to understand it. What we know is often based on preconceived ideas and personal experience, both of which prompt us to see human qualities in our pets and other animals (Shepard, 1978). Explaining animal behavior psychologically is widespread, and scientists are not immune to that trait (see Tolman, 1948; Menzel, 1974; Garcia, 1989). However, there is a difference between knowing animals in psychological or other ways, and scientific understanding of animals, whether psychologically or otherwise. Scientific understanding supersedes individual experience through systematic explanation of observable facts. To be scientific, an explanation must be valid and reliable. It follows that an explanation can be valid and reliable only when it predicts measurable and controllable novel phenomena (Brown and Ghiselli, 1950).

Understanding animals scientifically is difficult, but not impossible. Even though we can never know animals as they know themselves (Nagel, 1974), both laboratory and field research provide perspectives that are independent of the ubiquity of human views. It is a thesis of this book that research in naturalistic settings contributes novel information that is essential to a scientific understanding of animal behavior (see Hediger, 1955, 1964; Menzel, 1971).

Human Attitudes

Understanding animals has always been relevant to the survival and well-being of humans (Zeuner, 1963; Ucko and Dimley, 1969). Although the origins of human interest in the psychology of animals is lost in antiquity (Heiser, 1990; Hediger, 1964), it is clear that hunting and trapping, fishing, and the domestication of farm animals and pets has led to a wealth of psychological information that is of much value to humans (Bowman, 1977). Over the centuries, closer contact with animals bred personal attachment and easy anthropomorphic interpretation by humans of animals' needs and wishes. As a result, understanding animals often seemed to be simple.

This appearance, however, is deceptive because anthropomorphic assessment of animals varies widely and does not extend equally to all animals. As humans, we conceive of things as they appear to us. Dogs, cats, monkeys, and other primates seem to us to be more like humans than do snakes or rodents. However, what of them? How valid are humans' views of the natures of reptiles and rodents? More to the point, can we even agree on the psychology of primates, cats, and dogs, much less rodents and reptiles? Even so, rodents comprise, by far, the majority of all mammals and, as such, are certainly important to humans.

Animal Welfare

Public and scientific interest in animal welfare has risen to unprecedented heights. Since the days when animals held witness in English courts of law (Ritvo, 1987) and Magendie's discovery of the Bell-Magendie law of 1822 helped to usher in the first Society for Prevention of Cruelty to Animals in 1824 (Cranefield, 1974), public concern has prompted increasing legal constraint on the treatment of animals—research animals, in particular (Gallistel, 1978). For example, the latest revision of the Animal Welfare Act calls for regulation of the psychological well-being of a few animal species—notably, primates. Even so, there is concern that these new regulations will be interpreted so narrowly that critical species differences will be overlooked (see Snowdon in chapter 15). The likelihood of this occurring rests in the problem of defining animal well-being (Dawkins, 1980, 1984).

Humaneness to Well-Being

Because valid results are vital to good research, humane treatment of captive animals is of vital concern. Humane treatment has been evident in psychological research from its beginning. Even in its most behavioristic days, natural ways of behaving—that is, psychological well-being—were taken into account (Small, 1900a,b; Kline, 1898; see Galef, 1984; Timberlake, 1983). Even Skinner's rat box was adapted to the rat's way of feeding (Skinner, 1938).

However, such humane treatment is not obvious to the public. Animal well-being is. By definition, well-being refers to a condition of health, happiness, or prosperity. Happiness means a state of being pleased and content, with freedom from sadness, sorrow, and the like, with all these concepts being decidedly subjective in character. Humaneness is characterized by kindness, sympathy, tenderness, and mercifulness, with all of these traits being specifiable in terms of human actions.

Why, then, has emphasis shifted from definable criteria of animal welfare to subjective assessment of this welfare?

The influence of public opinion is not to be denied. Confusion about the substance of the issue abounds. Identification with pets—and other animals in our charge—makes the confusion readily understandable. However, in essence, the shift is due to other factors. Among them is the recent progress in describing behavior that is typical of particular species. In the absence of a clear definition of well-being, the public—and scientists alike—turns to what is normal in species' behavior in search of objective criteria.

The issue for science is humane treatment, not well-being. The goal of science is understanding the origins and causation of animal behavior. The delineation of normative species typical behavior is a necessary step in that direction. The development of new scientific approaches to animal behavior—such as sociobiology, or behavioral ecology—and the expansion of older ones—as ethology, or comparative psychology—has greatly accelerated the discovery of differences in normative behavior that clearly separate species as well as subspecies from each other. In addition, as a result of such advances in scientific knowledge, attention has become focused on the psychological aspects of animal life.

At the same time, the desire to completely understand animals forces experimental research into areas of neural and physiological functioning of living animals. To say the least, these animals may experience fear, pain, and discomfort. Thus, the juxtaposition of these two aspects of scientific research—both flowering in recent times—is largely responsible for the shift to emphasizing well-being and defining it as "freedom to perform behavior typical of the individual's species." In doing so, exercise of each individual's unique species heritage is acknowledged as being important to its well-being. However, there are shortcomings to this acknowledgment.

SPECIES-TYPICAL BEHAVIOR AND WELL-BEING

It is assumed that species-typical behavior is genetically based (Tinbergen, 1951; Price, 1984). If this is so, then the mode of life of a species links directly to its ancestral heritage. Modern animals were shaped by natural selection in the context of their ancestral habitats—habitats that have vanished or are disappearing or changing. Nevertheless, natural selection has not stopped. It continues even in captivity (Price, 1984, 1985). Therefore, a so-called normal environment for an animal need not involve the habitat or the niche in which its ancestors evolved.

As natural selection goes on, in either nature or captivity, current conceptions of species-typical behavior gradually become anachronistic. The wild animal of today must deal with changing habitats and shifting niche parameters, just as today's domesticated species must adapt to varied shifts and systematic changes in conditions of captivity. In the face of this change, selection favors differing individuals. As always, emphasis is on individual variation. Individually different forms of species-typical behavior must be emphasized, not only with a view to adapting existing animals to changing conditions in captivity, but with a view to assessing and understanding the role of experience in shaping psychological reactivity. The ideal of serving captive animals well might just focus on individuals, and how the behavior of each individual itself varies, rather than the norm of behavior of a species.

In any case, being well involves more than doing merely what is done in the wild. Even in nature, behavior alone cannot identify a state of well-being. Physiology and behavior change due to individual experience under both natural and captive conditions (Price, 1984). Experience of novel events and their repetition modify pituitary and adrenal activity and other measure of neuroendocrine actions often taken as indicative of psychological stressors (Levine, 1983, 1985). A valid definition as well as a reliable measurement of stress have yet to be discovered (but see Blanchard et al., 1989). Clearly the development of norms for a variety of health indexes—taken one animal at a time and in one captive population after another—remains a desideratum.

Given the ubiquity of stress in normal life and the adaptiveness of stressful reactions in dangerous situations—and apart from the difficulties involved in defining the concept—health indexes assessing neuroendocrine reactivity may never become particularly useful. Rather, the search should continue for norms of contextual conditions, physiological indicators, and behavior changes relating to major health detriments, such as loss of appetite; internal and external ulceration; overt injurious aggression; stereotypical fixed behavior; and self-injury. Again, the types and ranges of individual deviations should be emphasized.

NATURALISTIC SETTINGS AND SPECIES-TYPICAL BEHAVIOR

The continuum of research methods ranging from laboratory study to field research includes naturalistic settings as intermediate way stations, comprising anything from abstract representations of critical natural elements to realistic renderings of what is known about a particular animal's lifestyle. As such, laboratory settings blend into naturalistic ones. It is not always clear exactly where one method leaves off and the other begins.

Laboratory research is clearly naturalistic in intent and import. An experiment cannot be done without taking biological functions into account. In some cases, a biological system is forced to perform contrary to how it would act in nature. Nevertheless, the results can relate to how the system normally functions. In general, research seeks to preserve those relations between an animal and his environment so that specific questions may be answered. This is done by noting the variety of ways in which the bits and pieces of the animal's world—as they appear to us—can be put together. We know which combination of the bits and pieces available to us is effective only through systematic investigation into the ways in which an animal makes a living. How the animal adjusts to its life circumstances is a focal question. The answer to that question depends on our abilities to predict and control behaviors.

Prediction and control derives from asking, "What makes it happen? How does it develop? What does it do? How did it evolve?" (Simpson, 1962; Tinbergen, 1968; Mason and Lott, 1976). Good answers exhibit internal consistency or internal validity; deductive generalization or external validity; and generalization to nature or ecological validity. In essence, ecological validity depends on providing research settings that are representative of an animal's morphological, physiological, and psychological needs (Brunswik, 1951; Cook and Campbell, 1979; Neisser, 1976; Petrinovich, 1990).

The setting may be a realistic copy of an animal's natural habitat, but only if a reasonably complete description of the habitat as humans perceive it to be, is available. However, strictly copying realism from nature is unnecessary. Simple abstraction of certain natural variables is sufficient (Mason and Lott, 1976). In either case, however, the setting must be biologically and psychologically meaningful to the animal (Brunswik, 1955a,b; Tolman and Brunswik, 1935; Gibson, 1979).

Nature can be copied both indoors and outdoors. An outdoor setting permits exposure to a broader array of daily variations in natural conditions than is usually possible indoors, and it can sometimes be illuminating. For example, early experience in the outdoors proved to be critical to habitat-preference of a certain subspecies of deer mice.

In 1952, Harris let individual deer mice choose between artificial indoor habitats. One resembled the natural environment of a woodland variety (*Peromyscus maniculatus gracilis*), while the other simulated the environment of a grassland variety (*P. m. bairdi*). Preferences were true to type, whether the mice were laboratory-raised or caught in the wild. In studies published in 1963 and 1964, Wecker tested laboratory-reared descendants of Harris's grassland mice in a 16–by–100–foot outdoor enclosure, with half of its length extending from a bordering field into a woodlot. The mice showed no preference for either the field or woods. However, their pups were raised in 10–by–10–foot pens in the field for four to five weeks. When tested, they preferred the field. Early experience in the field had led to field-preference. When pups raised in wood-lot pens were tested, no preference for either field or woods appeared.

In nature, ground-dwelling rodents divide nest-building into two phases. First, they dig a burrow, and then they construct a nest. The burrow becomes a secure place—a center for foraging and other activities (Galef, 1982, 1985).

Riess raised female rats without opportunity to manipulate objects. As adults, and when housed in standard laboratory maternity cages, they did not build nests nor did they care for their newborn pups (Riess, 1950, 1954). In a study published in 1961, Eibl-Eibesfeldt raised female albino rats in the same way with the same result. However, when these rats gave birth in their own cages, with one corner divided by a small vertical partition, nest building and pup care were normal.

Observation of wild rodents has implicated a need not for the prior construction of a burrow or manipulatory experience, but for the consequences of burrowing—that is, the existence of a shelter—for rat the maternal behavior of rats to be normal. This proved to be true, even though, when offered earth in which to dig afterward, the rats did so. Nest-building is, for rodents, generally more dependent on external rather than internal factors (Glickman, 1973; Hogan and Roper, 1978).

This series of studies exemplifies the usefulness of observations made outdoors, whether in enclosures or in the wild. Attention was called to developmental factors or the importance of early experience as well as the frequency with which simple changes in captive settings—partitioning a cage into two separate areas can benefit an animal, even in highly constrained circumstances.

NATURALISTIC SETTINGS AND METHODS

What Is a Naturalistic Research Setting?

That animals typically occupy only a portion of the environment available to them is well-known. In nature, animals do not simply fit into niches provided for them. Indeed, the surroundings of an animal are defined by the animal itself. Animals actively participate in creating their own niches by transforming their environments. The roles which animals play in constructing their ecological niches from the bits and pieces of the world available to them is often overlooked.

In the wild, an animal effectively controls its environment in significant ways. It not only selects nest sites, refuges, and foraging times and places, but it also actively defends certain areas. In doing these things, it chooses the times and the extent of its movements, as well as the time, the type, and the degree of its social contacts. In addition, it seldom experiences conditions from which it cannot escape or alter by appropriate behavior (Kavanau, 1963).

Accordingly, a naturalistic setting should include representations of the bits and pieces of the animal's world of which we are aware, and experimental testing should be done within that setting. This implies that a setting should be an indoor or outdoor enclosure large enough to permit significant freedom of movement, in which the animal—or several of them—could, if necessary, live for prolonged periods.

Brant and Kavanau studied the behavior of wild (*Peromyscus crinitus, P. maniculatus*) and domestic mice (*Mus musculus*) with the concept of a home range in mind. The animals lived in enclosures so large that the size of the inhabited region was a variable under their control—that is, it depended on where they went and what they did. Equal access to all regions was provided by connecting cages, some empty ones and others containing food, water, dirt, an activity wheel, and a nest. The connections were made through a complex of burrow simulating mazes containing vertical and slanted passageways (Brent and Kavanau, 1964, 1965).

When placed in the enclosure, the mice behaved as if translocated in space. They oriented toward surfaces and objects just as animals do in the wild. Exploration of the enclosure, on a shorter time base, was like exploration in the wild. Movements from place to place resembled revisiting known parts of a habitat in the wild. Some of them nested at more than one place, as frequently happens in the wild.

In earlier work and using smaller enclosures, Brant and Kavanau found that mice quickly learned to operate seven manipulanda, all simultaneously active and involving four different functions. While they all operated motor-driven activity wheels, some preferred a square or triangular one. In rotating circular ones, they jumped over the axel (Kavanau and Brant, 1965; Kavanau, 1967a). They turned lights or sounds on and off, visited nests, and, after adapting, dimmed the light intensity to a preferred low level. The mice also learned "burrow-simulating mazes of unprecedented complexity—containing hundreds of blind alleys—without extrinsic reward" (Kavanau, 1964, p. 490).

Is Self-Control Important?

When the experimenter turned off a motor-driven wheel, the mice promptly turned it back on. When it was turned on after being off, the mice often turned it off. A dimmer-equipped light periodically turned on by an experimenter, was turned completely off, and if it was an illuminated light periodically turned off, turned fully on. Only after a period of adaptation did the mice dim the light, whether it was turned on or off, to a characteristically preferred low level (Kavanau, 1963). As pet owners know, animals are capable of and prefer self-control of their actions. Even the illusion of such independence frequently suffices (Hill, 1978; Lefcourt, 1973).

For example, the complexity of the factors involved is manifest in the behavior of threatened mother rats caring for their offspring. Lactating female rats, who experienced a threatening event only once, moved their pups into a neighboring compartment. They then moved their nest. Only after these moves did the female rats return to cover the source of the threatening event with available materials (Pinel and Mana, 1989).

Systematic research into the self-control phenomenon has examined animals' preferences for working at a task for a reward when the same reward is freely available (Osborne, 1977). Autoshaping is one example of this type of self-control, and possible explanations are varied and diverse. Stimulus-change related to response-contingent food delivery is one suggestion (Osborne, 1977). The exercise of foraging behavior, tracking, capture, and handling is another possibility (Collier, 1982). Exploratory tendency or curiosity could be yet another (Glickman and Scroges, 1966). Preference for control—because control modulates the affective value of an event—is still another (Overmier et al.,1980).

Affective modulation is seen in the importance of self-control to the proactive interference effect in learning avoidance and escape (Overmier and Seligman, 1967). The proactive interference effect—often interpreted as "learned helplessness"—is potentially compatible with changed activity level (Bracewell and Black, 1974), or altered motor response (Anisman and Waller, 1973). However they are explained, the effects of lack of control, loss of control, and the interaction of control and predictability—as relat-

ed to neurochemical and somatic effects, as well as behavior—
need to be dealt with (Levine, 1983; Mineka and Henderson, 1985).

ARE THERE UNIQUE NATURALISTIC RESEARCH METHODS?

In some respects, research in naturalistic settings goes
beyond traditional laboratory methods. In addition to facilitating
analysis of species-typical behavior, naturalistic methods expedite
discovery of the stimulating context in which behavior occurs.
Response-defined constructs, such as stress, are often diffuse and
ambiguous (Levine, 1985). Such constructs can be clarified by
analysis of relevant sensory input. Knowledge of perceptual cues is
vital.

Using a simultaneous discrimination apparatus, in a study
published in 1938, Krechevsky found proximity and closure to be
critical in rats' preference for vertical over horizontal displays.
Menzel (1969), using observational methods, also found vertical
extent to be critically important, but this time in monkeys' perception of objects in the field. His data were collected with little or
no disturbance of his animals' normal daily routine. Krechevsky
could have done the same with his rats if a suitable naturalistic
setting had been used (for example, compare Thompson and
Solomon, 1954; Gibson and Walk, 1956).

Habituation of freely moving animals in naturalistic enclosures offers atypical ways of assessing animal preferences. Aspects
of the setting can be altered or novel features can be introduced to
test for critical perceptual elements. In both cases, habituation can
be measured. If a change is unimportant, then response is unlikely
or responsiveness declines quickly with little recovery. If a change
is important, responsiveness is greater and perhaps qualitatively
different, and the subjects recover repeatedly on retesting (Wyers
et al., 1973; Menzel, 1964; Menzel and Menzel, 1979).

Use of this method does not exclude choice-testing, but
bypasses some of the problems of assessing perceptual reference by
that method (Curtis, 1985). Choice-testing usually involves a limited number of selections among a few alternatives in a structured
situation (Dawkins, 1983; Krechevsky, 1938). Useful information
is often gained, but, as Curtis (1985, p. 10) has noted, "not only is
previous experience critical, but the form of the test itself affects
results, especially pertaining to the choices made available. Alter-

natives important to well-being may seldom be chosen and animals may not always choose what is best for them."

There can be no denying the difficulties noted. However, they do not denigrate the usefulness of evaluating animal perceptual orientation. Observation of freely moving animals in naturalistic settings—as in nature—provides other ways of evaluating preferences. Rather than forcing a choice, movements toward or away from an altered feature can be used to evaluate preference, stress, or indifference. This type of test requires naturalistic enclosures. It is important that the animal subject be free to move about, as well as control where it is and what it does in its habitat enclosure.

How large should the enclosure be? It must be at least large enough to permit the display of withdrawal, defensive, courting or other appetitive or foraging behavior. What sorts of novel alterations and changes in the structure and routine of the setting should be introduced? Look to nature, both for the wild and the domesticated animals for guidance. Make changes designed to answer causative, developmental, functional, and adaptive questions relating to the species of concern. Over time, make repeated and varied changes against which to measure novelty and habituation. In short, test experimental hypotheses within the enclosure by altering the setting as Menzel has done (Menzel, 1984; Menzel and Juno, 1985).

If movement patterning or motor control is of interest, unique instrumental manipulanda requiring particular modes of response can be introduced into the setting after the manner of Markowitz (Markowitz and Stevens, 1978; Markowitz, 1982). For example, Bolles reported rats as having great difficulty learning to do two things with one manipulandum. When they learned to do so, they did each thing using a motorically different response pattern (Bolles, 1989).

The features of any setting create unique demand - characteristics which an animal must express. Thus, needs exist in every setting that go beyond basic physiological processes. These, too, must be acknowledged by animals and observers alike. Our question is whether those needs carry over to other modified settings and nature, such as seen in the ecological validity question. Hence, the need for varied and extensive naturalistic settings is paramount.

How well do the heritages of captive and domesticated animals serve them in the wild? Feralization studies provide an answer and, here, naturalistic enclosures provide possibly the only setting of choice (Boice, 1981).

White laboratory rats did very well when left on their own in a large outdoor enclosure. Each of five pairs of rats, which were provided with familiar nest boxes, buried with the top surfaces flush to ground level, promptly left them to begin digging burrows. The rats extended their burrows after the manner of wild rats (Lore and Flannelly, 1977), and they were successful in raising offspring. All ten survived their first winter and, over a two-year period, increased in number to around fifty. During this time little food was hoarded and serious fighting was probably absent (Boice, 1977). Although feralization was not complete, marked increases in wild-type behaviors appeared as soon as in the first generation born outdoors (Clark and Galef, 1977).

Even monkeys which were socially maladapted through early experimental deprivation (Harlow and Harlow, 1962, 1969) rapidly readjusted to normal social behavior when allowed freedom of action as a social group (see Novak et al., in chapter 16). When given freedom, laboratory quail (*Coturnix coturnix japonica*) exhibit normal parental behavior (Hess et al., 1976), and domestic Peking ducks display social behavior typical of mallard ducks, *Anas Playtyrhynchos* (Miller, 1977a).

Apparently, and in spite of selective genetic changes, domestication does not necessarily result in behavioral degeneracy (Boice, 1981; Price, 1984). Major genotypical change due to domestication alone is rare (Berry, 1969; Bader, 1956). The changes induced are, for the most part, selectively related to experiential developmental alternatives. Thus, domesticated and captive animals may be more like their wild brethren than is often thought (Price, 1984; Boice, 1973, 1981). Such results reflect psychological resiliency in the face of captive conditions denying opportunity to exercise species-typical behavior. Does this mean that barring compulsive and self-injurious behavior (including injurious escalated fighting), typical captive environments serve the well-being of research animals well? For the white rat—not to mention the quail and duck results also noted—laboratorization produced effects that quickly rubbed off in the wild.

CONCLUSIONS

The well-being of research animals is an imperative for accurate description and interpretation of behavior. The conference giving rise to this book subscribed to the theme that, even though we can never look at the world through an animal's eyes, the best way to learn how to scientifically understand animals is through observing and measuring their behavior in as wide an array of research settings as possible. We need to know what aspects and properties of objects—as we see them, and as they are distributed in space—influence and control the behavior of an animal under study. Even if we could envision the world from the animals' viewpoint, we would not be thus enabled to make a true assessment of its behavioral state.

What we can do is observe behavior from as narrow or as broad a physicalistic perspective as is required, sometimes "zooming in" for a close-up and at other times "zooming out" for a view of wider scope (Menzel, 1969; Menzel and Wyers, 1981). By doing so, general observation, becomes not just supplemental to, but a counterpart of experimentation (Menzel, 1969, p. 81). When we do so, we gain a more complete view of animals as a whole—but it is always *our* view and not *theirs*.

From our view, the use of naturalistic settings provides many direct benefits to science. Information is made available for designing more efficient research settings by consideration of questions such as the following: Which pieces of the world do animals find relevant? How do animals perceive and utilize these pieces during the course of their lives? How do relevant components of the environment relate to each other? The answers to these questions permit construction of abstract models of nature within which the essential elements of the ecology can be assessed. The physical environment provides habitats and ecological niches, but niches cannot exist without animals. The animal is a necessary part of the definition of a niche (Gibson, 1979). Thus, naturalistic settings allow us to generalize to nature, as well as to establish ecological validity and learn what animal nature is about. Does research in naturalistic settings aid and abet debate over the welfare of captive research animals? Certainly so—and for a variety of reasons.

1. The research at least suggests animals are living as they do in nature.
2. Criteria for assessing the psychological well-being of animals are thereby more easily imagined.
3. The research promotes ideas concerning the redesign of captive animal housing.

4. Such ideas may well meet public concern about research animal housing, as well as psychological well-being.
5. New ways are fostered for veterinarians to assess animal health and physiological condition.

Simulation of the effects of behavior can benefit animal subjects even in highly constrained but representative circumstances. It is possible to approach federally mandated requirements (Irving, 1985; see also Annelli and Mandrell, in chapter 5; and Bayne and Henrickson, in chapter 4) by outfitting naturalistic research settings with materials other than those ready-made by nature (see also Doherty and Gibbons, chapter 10, and other contributions within this volume).

The concerns that motivate regulation of animal research are easily recognized. We who espouse the use of naturalistic settings are necessarily concerned with the welfare and well-being of our animal subjects. It is easy to understand how many people feel about the well-being of research animals, and we want to do what has to be done. The problem for us lies in arbitrary requirements and procedures that are not based on research data. Cosmetic changes that look good to humans, but do not benefit captive animals, are a disservice to science and to the animals themselves.

The range of views regarding the use of animals encompasses an epistemological puzzle. What is an animal? Is it the individual that a human has to cope with at a particular time? Or is it the species of which that individual is representative? To ask which individual or which species is a cogent question. It is my view that, as scientists concerned with understanding the processes of nature as a whole, we must answer to the individual human addressing an individual animal.

II. ANIMAL CARE AND USE:
REGULATIONS AND CASE HISTORY

KATHRYN BAYNE
ROY HENRICKSON

4

Regulations and Guidelines Applicable to Animals Maintained in Indoor Seminaturalistic Facilities

The tradition of housing wild animals used in biomedical or behavioral research in captive naturalistic settings has long-standing roots. The zoo community has aptly demonstrated the feasibility, and indeed the benefits, of maintaining animals in complex habitats while still meeting United States Department of Agriculture (USDA) standards of animal care. In fact, the trend of the last decade has been to "naturalize" the older, more sterile, exhibits. Thus, zoo housing has shifted from bar and tile cages to elaborate replications of the animals' home range. This concept has been carried over to laboratory facilities. For example, regional primate breeding centers tend to maintain much of their large populations of breeding monkeys and apes in outdoor compounds. Certainly this situation represents the most extreme case of naturalizing the laboratory animal's habitat when compared to laboratory situations whereby natural components are included in indoor housing conditions. Yet these centers have struck a harmonious balance between maximizing animal reproduction, providing an appropriately social environment, and maintaining animal health. Also, scientific studies probing means of providing for environmental enrichment suggest that increased environmental complexity contributes positively to the animals' well-being (e.g., Bayne et al., 1991; Schapiro et al., 1991; Westergaard and Lindquist, 1987).

Many laboratory facilities have also explored using naturalistic conditions as a potential routine manner of indoor and outdoor animal holding. Laboratory facilities, however, must accommodate several regulatory standards in addition to those of the USDA. These guidelines, standards, rules, and regulations are directed primarily toward traditional laboratory species. They can pose special problems for seminaturalistic schemes of housing. Factors to be considered in designing or approving indoor seminaturalistic housing include the unique requirements of the species being studied, the research objectives, and the applicable rules and regulations. Conflict arises when justification is sought for poorly designed housing erected in substandard facilities. Examples are chicken wire and untreated wood enclosures in rooms lacking environmental controls or incapable of sanitation. Conflict may also arise when the requirement for all stainless steel sanitizable enclosures is imposed. Such caging may preclude the addition of cage enrichment that may be required for behavioral expression, reproduction, and fulfillment of research objectives.

Although this chapter addresses indoor naturalistic habitats, many of the principles that will be discussed for indoor habitats are applicable to outdoor naturalistic environments. Problems encountered with outdoor facilities include the protection of animals from inclement weather and the use of natural components for cage enrichment. If tropical animals are kept outdoors in temperate climates, supplemental heat may be necessary. This creates a number of problems requiring unique strategies. Cage enrichment using natural materials such as log jungle gyms are sometimes problematic with regulatory agencies because of the inability to properly sanitize surfaces. If animal densities are kept low and groups of animals reside in the same cage for long periods, concerns about sanitation may be outweighed by the benefits of the cage enrichment.

The working premise of this chapter is that facilities can be designed with indoor naturalistic holding conditions in such a way that they are not in conflict with these standards. Any disparity between the design of naturalistic facilities and existing regulations and guidelines can be avoided by examining the facility, and making the appropriate adjustments based on a two-prong approach: (1) a critical assessment of the physical plant, and (2) programatic review.

THE RULES AND REGULATIONS REGARDING ANIMAL CARE AND USE

During the past five years many of the rules, regulations, and guidelines governing the care and use of laboratory animals used in biomedical and behavioral research have been revised. Those revisions have placed increased demands on institutions for central monitoring of animal care and use programs, for improving animal husbandry practices and housing, and for providing increased veterinary care. These increased demands have caused concern among scientists, particularly those involved in research with animals kept in indoor naturalistic environments. Issues such as management control, cage construction, and sanitation practices are central to these concerns.

The Animal Welfare Act

Regulations regarding the Animal Welfare Act (AWA) of 1966 (P.L. 89–544) have been amended three times, most recently in 1985. Subpart D of the AWA contains specifications for the humane handling, care, treatment, and transportation of animals. Most relevant to discussions of animal holding are sections 3.75–3.81. In these sections, standards for both physical plant and program are established. The structural strength of the facility, provision of adequate water and electric supply, waste disposal, pest control, and the establishment of programs of disease management, euthanasia, and veterinary care are just a few of the common considerations that must be given to all animal facilities under the AWA.

Individuals who utilize indoor naturalistic environments for their laboratory animals must pay particular attention to the requirements that mandate sanitization of cages, rooms, and hard-surfaced pens or runs by a cage wash system, steam cleaning system, disinfection process every two weeks. The AWA also mandates the removal of soiled gravel, sand, or dirt, and replacement with clean substrate when necessary. This latter requirement is stated in general terms, and thus allows for inclusion of professional judgement in the determination of a cleaning schedule.

However, a significant difference in the management of naturalistic outdoor and naturalistic indoor facilities clearly exists. Although consideration for the elements can complicate management of outdoor habitats, factors of ultraviolet lighting, desiccation by wind, and dilution by rain may be considered beneficial processes for maintaining sanitary conditions. As these elements are not at work in indoor naturalistic environments, other considerations must be given to the cleaning of these facilities.

Public Health Service Policy on Humane Care and Use of Laboratory Animals

The Health Research Extension Act of 1985 (PL 99–158) mandated the already existing Public Health Service (PHS) Policy on Humane Care and Use of Laboratory Animals, with only minor modifications as reflected in the 1986 revision. The PHS Policy establishes guidelines for activities involving animals used in biomedical and behavioral research conducted or supported by the Public Health Service. In addition, the PHS Policy requires the establishment of Institutional Animal Care and Use Committees (IACUCs) to act as agents of the institution. The IACUC must be composed of no less than five members, one of whom must be an appropriately qualified veterinarian with direct or delegated program responsibility for activities involving animals at the institution. Finally, as a condition for receiving PHS support, institutions are required to file Animal Welfare Assurances that are satisfactory to the Office for Protection from Research Risks (OPRR), guaranteeing that these guidelines are being met.

Those provisions of the PHS Policy that particularly impact laboratory animal programs with naturalistic housing conditions include: (1) the provision that the IACUC must evaluate and report on animal facilities and programs semiannually using the *Guide for the Care and Use of Laboratory Animals (Guide)* as a basis for evaluation; (2) the provision that the IACUC must review and approve research projects and confirm that they will be conducted in accordance with the Animal Welfare Act, the *Guide,* and the institutional Assurance; and (3) the provision that training of laboratory personnel on many topics including humane animal care will be conducted.

As is the case for most of the regulations pertaining to laboratory animal care and use, the language of PHS Policy is quite general. There are no specific standards in the Policy that would preclude naturalistic housing. On the contrary, section IV.C.1.d. mandates that:

> The living conditions of animals will be appropriate for their species and contribute to their health and comfort. The housing, feeding, and nonmedical care of the animals will be directed by a veterinarian or other scientist trained and experienced in the proper care, handling, and use of the species being maintained or studied.

Like the Animal Welfare Act the policy requires adequate medical care provided by a qualified veterinarian.

In PHS Policy, the IACUC is made responsible for semiannual review of the institution's program for humane care and use of animals and semiannual inspection of the institution's animal facilities, using the *Guide* as the evaluation standard. As the IACUC membership is composed of individuals with an understanding and expertise in the specific research area under discussion or in laboratory animal science/medicine, well-designed naturalistic habitats that conform with current regulations and guidelines should be considered acceptable by the IACUC.

The Guide for the Care and Use of Laboratory Animals (Guide)

The Guide (1985 revision), a publication of the National Institutes of Health (NIH), defines the critical elements of an institutional animal care and use program. The *Guide*, a relatively detailed rendering of animal care and use guidelines, was prepared to "assist institutions in caring for and using laboratory animals in ways judged to be professionally and humanely appropriate" (*Guide*, Preface). Compliance with the *Guide* is required of any institution receiving funds from the PHS. The *Guide* is written in general terms so the recommendations can be applied to a variety of institutions using animals in research and testing. Because it is written in general terms, "professional judgement is essential" (*Guide*, p. 2) in the application of the recommendations.

The *Guide* provides recommendations on institutional policies to be followed in establishing an animal care and use program.

Central to these recommendations is the establishment of an animal care and use committee that reviews and monitors the program. The remainder of the *Guide* covers in detail animal husbandry, veterinary care, and facility design and construction. These guidelines apply primarily to common laboratory species kept under standard conditions. Naturalistic facilities have unique requirements based upon the species kept and the research objectives. Providing that the basic principles recommended in the *Guide*, such as daily monitoring of animals, veterinary care, proper sanitation, and the like are adhered to, the *Guide* does not preclude the keeping of animals in naturalistic settings.

Fundamental to the recommendations in the *Guide* is concern for the well-being of the animals. Animals can be kept successfully in naturalistic environments provided such use is justified scientifically and approved by the institutional animal care and use committee. It is essential that the institutional veterinarian be actively involved in the planning and management of such facilities.

As the *Guide* is the cornerstone of the regulatory standards that facilities must meet, a close examination of those parts of this document that directly apply to indoor naturalistic housing conditions is in order. This examination will be composed of a "walk through" each section of the *Guide*.

Introduction. Perhaps one of the most revealing statements regarding the intent of the *Guide* is contained in the introduction. It is made clear in this section that these guidelines are not intended to restrict an investigator's freedom to pursue animal experimentation as long as scientific and humane principles are followed. In addition, it is stated that improved techniques in laboratory animal care and use may be encouraged by application of the *Guide* by scientists. If it can be demonstrated that a naturalistic setting is indeed a better way to house a species, then by *Guide* recommendation such a method should be implemented.

Two poignant examples of this occurred at the University of California, Berkeley. The first case involved the housing strategy for Jamaican bats. Husbandry practices were dramatically altered from standard laboratory animal conditions to keep the bats in good health. Temperatures were maintained at 30°–33° C, humidity at 90% and the air flow was reduced to approximately one change per hour. In addition, the use of sanitizing agents was greatly restricted, because of the sensitivity of these animals to chemical agents. The animal room was converted to a dank, dark smelly bat cave, and the animals thrived.

The second case involved the management of kangaroo rats. Kangaroo rats are nocturnal burrowing animals requiring a seed diet and sand substrate (for sand bathing). Keeping this kind of animal in a clear plastic shoebox with sawdust bedding and providing commercial rodent diet did not provide a proper environment. These animals were provided with a sand substrate, a seed diet, and a burrowlike darkened area where they could seek refuge.

Both these cases underscore the most important statement in the *Guide*: "Professional judgement is essential in the application of these guidelines" (*Guide*, p. 2).

Institutional Policies. A description of the mission of the Animal Care and Use Committee (IACUC) for an institution is provided in this section. The IACUC is primarily responsible for programatic issues. According to the *Guide*, "knowledge of the husbandry needs of each species and the special requirements of research . . . programs. . . ." (*Guide*, p. 3) are essential to a successful IACUC. Clearly, an identifiable need for unusual housing conditions of an animal should be acceptable by the IACUC.

Institutional policies, according to the *Guide*, should make provision for adequate veterinary care, qualified care personnel, appropriate personnel hygiene, and an occupational health program. This type of program is not incongruous with a well-managed naturalistic animal holding facility. One key factor in the successful management of these support programs is clear lines of authority and responsibility. Also, the IACUC can aid the institution in developing policies that will streamline the management of all the animal programs. This committee must evaluate and monitor the program for compliance. Concerns that center around the types of materials permissible for cage construction and enrichment, the schedules for sanitation, and overall management should be reviewed by the IACUC. The peer and veterinary resources available through this committee should be tapped.

Laboratory Animal Husbandry. As the *Guide* (p. 11) states, "a good husbandry program provides a system of housing and care that permits animals to grow, mature, reproduce, and maintain good health." This assertion leaves much room for innovative housing strategies, so long as the general well-being of the animals is promulgated. The authors of the *Guide* were aware that facility management must be customized due to the variability in both "subjective" and "objective" factors (p. 11) operating within each institution. Indeed, the *Guide* (p. 11) makes it clear that institutions with "less than optimal" facilities can still provide quality animal care through the dedication of well-trained and motivated personnel.

The caging or housing method is cited as being one of the most critical components of the physical and behavioral environment for laboratory animals. Several recommendations are made in the *Guide* as to what constitutes a suitable housing system. Salient among these recommendations is that the housing method provide a comfortable environment. This necessarily dictates an understanding of the biological and behavioral needs of a species (for example, the provision of branches to marmosets for scent marking). If the naturalistic housing system is to meet these guidelines, it must (1) be escape-proof, (2) provide ready access to food and water, (3) have adequate ventilation, (4) be appropriate for the biological needs of the animal, and (5) protect the animals from known hazards. A well-designed naturalistic habitat should have no trouble in meeting these guidelines. Examples of naturalizing the laboratory environment currently in use include the provision of artificial fleece or turf foraging substrates for primates (Bayne et al., 1991, 1992), swimming opportunities for captive beaver (Gilbert, personal communication, 1987), and nestbuilding substrates for cotton rats (Chedester, personal communication, 1992).

According to the *Guide*, caging methodologies should effectuate research while maximizing the health of the research animals. The cages should "promote physical comfort" (p. 12) of the animals. They should be designed with care so as to minimize risk to the animals (e.g., minimize spread of infection between cages, and be maintained in a state of good repair), ". . . facilitate animal well-being. . . ." (p.11), accommodate requirements of the research, and minimize experimental variables. A strict interpretation of the *Guide*, then, would suggest that every effort should be made to minimize stress that results from the housing environment itself and that some degree of flexibility in housing design elements should be promulgated.

Where appropriate, the *Guide* says, ". . . group housing should be considered for communal animals" (p. 12). When providing a social environment, factors such as degree of familiarity between animals slated for introduction, optimal animal density and composition, and the ability to escape from other animals should be taken into account. In addition, it is recommended that environmental enrichment suitable for the research animal be provided, particularly for animals used in long-term studies.

When considering space recommendations, the *Guide* provides specific minimum cage sizes, but also provides general advice. For example, ". . . successful experience and professional judgment" (p. 13) are deemed to be requisite as ". . . it is unlikely that a single ideal or perfect system. . . ." of caging can be identified (p. 13). In fact, the *Guide* states that "special housing provisions are sometimes necessary for unusual laboratory species such as those with unique metabolic or genetic characteristics or special behavioral or reproductive requirements" (p. 13). The comment in the *Guide* that might be considered the bottom-line recommendation regarding the animal environment is that "the animal environment in which animals are held should be appropriate to the species and its life history" (pp. 17–18).

The *Guide* sets forth recommendations for some of the specific components of an animal's environment. Guidelines for suitable temperature/humidity, ventilation, illumination, noise, water, and sanitation are among those discussed. From a researcher's perspective these factors are critical for the success of an experiment for many reasons. For example, "temperature and humidity are probably the two most important factors in an animal's physical environment because they can affect metabolism and behavior" (p. 18). And what study is not related in some manner to either or both of these elements of an animal's makeup? Ventilation standards can have an effect on an animal's well-being by virtue of the ammonia accumulation that can occur in a cage. Not only can this adversely affect the animal's health, but can prove to be a confounding variable in an experiment. Light levels can also affect animal maintenance. For example, light periodicity can impact reproductive and endogenous cycles. Even the bedding used can result in physiological changes in research animals, such as the changes in microsomal oxidative enzyme levels recorded in rodents housed on pine or cedar shavings (Weichbrod et al., 1988).

Other animal husbandry issues noted in the *Guide*, which can be readily accommodated in indoor naturalistic housing, include noise control, access to fresh, potable water, and a sanitation program that includes general cleanliness, waste disposal, and vermin control. Also a program of animal identification should be instituted for better animal tracking and improved historical background on the research animal. Provision in the animal care and use program for emergency, weekend, and holiday animal care and adequate veterinary care should be established.

Physical Plant. A well-designed physical plant is a critical component to achieving a well-managed animal facility. The *Guide* makes it clear that "the design and size of an animal facility depend on the scope of institutional research activities, animals to be housed, physical relationship to the rest of the institution, and geographic location. A well-planned, properly maintained facility is an important element in good animal care. The animal facility must be designed and constructed in accordance with all applicable state and local building codes" (p. 41).

The *Guide* recommends that within an animal facility there be animal holding and support space. The animal holding should provide sufficient space for three main activities: (1) adequate separation of species or research projects as necessary, (2) quarantine and isolation procedures, and (3) animal holding. Throughout the facility, recommended construction guidelines include building materials that enhance ". . . efficient and hygienic . . ." (p. 43) operations.

Several of the important physical plant standards that are provided in the *Guide* can readily be accommodated in the design and construction of indoor naturalistic animal holding. These include floors, walls, and ceilings that are sealed against moisture, wear, and chemicals. The protection of walls from damage resulting from movable equipment can be incorporated into and masked by the naturalistic design. Both the walls and ceilings should be constructed with impervious junctions.

Special Considerations—Farm Animals. In many ways the management considerations of farm animals used in biomedical research overlap with those of other laboratory animals housed in naturalistic conditions. For this reason, an examination of the comments in the *Guide* that pertain to farm animal housing practices is worthwhile. In this section of the *Guide*, it is again stressed that ". . . housing and management practices should be designed to provide optimal animal care" (p. 51). It is suggested that the index for optimal housing be animal well-being, a concept that has a broader meaning than survival and production.

Although climatic control is not as much an issue in an indoor holding room as it might be outdoors, maintaining the animals in their ". . . thermocomfort zone . . ." is recommended for ". . . maximum performance and physiological stability . . ." (p. 52). It is noted in the *Guide* that the appropriate environment will vary with species and age of the animal, and will depend upon research goals.

Facility construction guidelines for farm animals contain a few relevant points for naturalistic housing. For farm animals, wood can be ". . . satisfactory . . ." if it is coated and sealed in a manner suitable to allow disinfection of the surface. In addition, animals that are fed in groups should be provided with enough separate feeding sites to minimize competition for resources. Water should be accessible to animals of all ages and physical condition, and should be provided in adequate supply.

Guide for the Care and Use of Agricultural Animals in Agricultural Research and Teaching (Agricultural Animal Guide)

Pursuing the theme that housing standards for farm animals may be relevant to indoor naturalistic conditions, a review of the *Agricultural Animal Guide* provides much of the same information contained in the NIH *Guide* (for laboratory animals). There are, however, some statements that bear special attention. For example, in chapter 2, General Guidelines for Animal Husbandry, it is noted that ". . . an ethically acceptable level of animal welfare exists over a range of conditions provided by a variety of . . . systems, not just in one ideal set of circumstances" (p. 7). Indeed, "proper animal management is essential to the well-being of animals, validity of the research, effectiveness of the teaching, and health and safety of animal care personnel" (p. 6). These points are perhaps the most representative of the position of this paper. In addition, the *Agricultural Animal Guide* states that an animal's well-being is dependent upon a variety of environmental components. Considering the complexity of the environmental matrix, then, it may be proposed that a "cookie-cutter" approach to facility design may not always achieve the optimum set of environmental circumstances for all species. This guide proposes that animal well-being can best be monitored by evaluating four factors: ". . . (1) reproductive and productive performance, (2) pathological and

immunological traits, (3) physiological and biochemical characteristics, and (4) behavioral patterns" (p. 6).

CONCLUSIONS

The provision of indoor naturalistic habitats to certain
species of laboratory animals can be problematic. However, strict
interpretation of the rules, regulations, and guidelines applicable
to animal care and use do not specifically preclude the use of these
types of housing conditions. Rather, a housing system that opti-
mizes the animal's well-being, and which is in accord with generally acceptable husbandry standards, and with the research goals is
encouraged.

Central to the success of any animal program is the establishment of an animal care and use committee that will review and
monitor the program. As the rules, regulations, and guidelines per-
tinent to animal care and use evolve, adjustments must be made in
the program to accommodate changes in standards. However, the
fundamental programatic and physical plant issues with which an
animal facility must be in compliance remain essentially the
same. Using the principles outlined in the Animal Welfare Act,
PHS Policy and the *Guide*, as well as professional judgment, naturalistic animal holding areas can be designed which are humane
and promulgate well-being .

JOSEPH F. ANNELLI
TIMOTHY D. MANDRELL

5

USDA Inspection Procedures: Nonstandard Facilities, Site Visits, Waivers, and Enforcement

When humankind first began to domesticate wild animals, it took on the responsibility to care for them. Their every need had to be provided for and met, or they would die. As humankind evolved, it kept its animals with it and used them as it needed. Aristotle, around 384–258 B.C., may have been the first to use animals in experimentation by dissecting and reporting internal differences between animals (Fox et al., 1984). Society also evolved and imposed standards of behavior for humankind. These standards also included the way in which humans treat the animals in their care. As society changes, so will the way in which humans care for animals both in captivity and in the wild.

In the United Kingdom the Cruelty to Animals Act of 1876 was passed to protect animals from abuse. In the United States various state governments passed local humane laws, but it was

49

not until 1966 that the first federal law to protect animals was passed. At that time Congress considered and passed the legislation that is known today as the Animal Welfare Act. This action was in part a response to articles in *Sports Illustrated* on November 27, 1965, and *Life Magazine* on February 4, 1966, about the misuse of animals in research. The Laboratory Animal Welfare Act of 1966 (public law 89–544) was passed primarily for the protection of private citizens' pets (dogs and cats) from possible theft and use in biomedical research (Fox et al., 1984). The Act covered researchers and dealers supplying domestic dogs and cats, nonhuman primates, rabbits, hamsters, and guinea pigs to research facilities. As written it was to protect dogs and cats being held for use in research. The key word in the legislation was "held" as this led to a narrow intrepretation of the Act and covered animals only before and after their use in research. The problem this posed was one of enforcement. A facility could claim that all dogs or cats housed were under study and therefore not subject to inspection.

As society has changed, so has the legislation and amendments to the Act. These changes have caused the Act to be modified several times. Each amendment to the Act has added new and stronger controls on the use of animals. The Act was amended in 1970 and given the name by which we know it today, the Animal Welfare Act (AWA). This amendment broadened the Act's scope to include all warm-blooded mammals with the exclusion of horses not used for research and other farm animals used for food or fiber. This 1970 amendment also expanded the number and type of facilities that were subject to the Act. The most significant inclusion involved organizations, such as zoos, that exhibit animals. The scope of the AWA was again expanded in 1976 to include transportation of animals by common carriers, such as airlines. At that time regulations were written to include marine mammals. The most recent amendments to the AWA were made by Congress in 1985. A few of the significant changes were the requirements for an Institutional Animal Care and Use Committee, requirements for reducing or eliminating pain and distress in laboratory animals, and requirements for the exercise of dogs and an environment adequate to promote the psychological well-being of nonhuman primates. On April 5, 1990, the Secretary of Agriculture also chose to begin regulating horses and other farm animals used in biomedical research.

As far back as 1966 Congress has entrusted the Secretary of Agriculture to enforce the AWA. The Secretary of Agriculture has given that responsibility to the Regulatory Enforcement and Animal Care unit of the Animal and Plant Health Inspection Service (APHIS). As APHIS is the agency within the USDA that employs veterinarians, it was the logical choice to administer the program. To be able to understand the way in which veterinarians working for a government agency enforce an act of Congress we need to compare the wording of the AWA as written and passed by Congress and the way the regulations are written and enforced by the Department of Agriculture. As an example of the difficulty in understanding and interpreting an act of Congress let us look at what would otherwise be a simple concept, the term "animal." As covered in the present version of the AWA, an animal is defined as:

> Sec. 2. (g) . . . any live or dead dog, cat, monkey (nonhuman primate), guinea pig, hamster, rabbit, or such other warm-blooded animal, as the Secretary may determine is being used, or is intended for use, for research, testing, experimentation, or exhibition purposes or as a pet; but such term excludes horses not used for research purposes and other farm animals, such as, but not limited to livestock and poultry, used or intended for use as food or fiber, or livestock or poultry used or intended for improving animal nutrition, breeding, management or production efficiency, or for improving the quality of food or fiber. With respect to a dog the term means all dogs including those for hunting, security, or breeding purposes; . . .

The AWA as passed by Congress can be a general document that sets out to express the will or intent of Congress. However, there are some very specific guidelines that Congress includes because it sees these items as crucial to the legislation, but in many areas it is somewhat vague in meaning. Congress, not being expert in the field of animal care, simply provides the framework for the construction of regulations. The regulations, which are written by APHIS, carry out the will or intent of Congress in a way that is understandable, measurable, and enforceable. Before publication, any proposed regulations must be reviewed by the department's Office of General Counsel to ensure that they are within the realm of the law and enforceable. Proposed regulations are also reviewed by the President's Office of Management and Budget for their potential financial impact on the federal government and the regulated community. To illustrate the differences and similarities between the AWA and the regulations, let us compare a section of the Act with the regulation. First, a portion of section 13 of the Act as amended in 1985: "Section 13(b)(3). The Committee shall inspect at least semiannually all animal facilities." Now, a portion of the regulations based on section 13(b)(3) of the Act: "Sec. 2.31 (c) (2). Inspect, at least once every 6 months, all of the research facility's animal facilities."

While the AWA and regulations are similar, the regulations are more specific. This specificity of meaning is provided in measurable terms that make the regulation a legally enforceable document. While it may be argued that some of what is expressed in the regulations was never in the Act, Congress allowed for this by including section 21 of the Act: "Section 21. The Secretary is authorized to promulgate such rules, regulations, and orders as he may deem necessary in order to effectuate the purpose of this act."

As the breadth and scope of the amendments increased, so did the quality of care. The trend over the last 10 years has been not just to provide an adequate living space but an enriched one. At one time it was considered sufficient to provide water, food, and a place to lie down. Today, not only are these items essential but the quality of the enclosure is equally important. Originally, the AWA considered only the prevention of theft and use in research, then later the medical and physical well-being of the species being housed, and today it also includes the psychological well-being of animals. Regulations for the exercise of dogs and an environment to promote the psychological well-being of nonhuman primates were finalized February 15, 1991. A few of the significant differences between the AWA as amended in 1976 and the one amended in 1985 are listed in Table 5.1.

While the AWA is the only enforceable law that legally covers the care and use of animals in exhibition or transportation, there is an additional law for animals used in research. The 1985 Health Research Extension Act requires that all institutions receiving Public Health Service (PHS) support comply with the PHS Policy on Humane Care and Use of Laboratory Animals. Additionally, the PHS policy requires that research facilities receiving PHS support comply with the National Institutes of Health's *Guide for the Care and Use of Laboratory Animals* (see Bayne and Henrickson, chap. 4, this volume). Further information on the PHS policy can be obtained from the Office of Protection from Research Risks (OPRR), Office of the Director, National Institutes of Health, Bethesda, MD 20205.

Compliance with the standards of the AWA is of concern to not just a regulatory agency but also to the public and research communities. With any law or regulation, to be effective it must be enforced. One method the United States Department of Agriculture uses to assess compliance with the AWA and related regulations is through regular and thorough site visits. The site visit is intended not just to regulate but to educate. Visits are unannounced and occur as frequently as necessary to insure compliance. As stipulated in the Act this is to be no fewer than once a year. In previous years the USDA has conducted an average of 1.6 visits per site at all research facilities, but more frequent inspections occur at facilities with the potential for problems. During a site visit, the inspector evaluates every aspect of the facility and its animal care program for compliance with the regulations. Items under the inspector's scrutiny are of four categories: (1) records, (2) facilities and operating standards, (3) animal health and husbandry, and (4) transportation standards, if applicable. The method that is used to document the findings is a written inspection report. This report addresses each section of the regulations (Figure 5.1). When APHIS inspectors discover noncompliance to regulations during an inspection, the items are documented on the inspection report and the necessary corrective measures are discussed with the official of the facility. The completed report is signed by the APHIS inspector and a representative of the facility.

Noncompliance is the term used for any portion of a facility or animal care program that does not meet regulatory requirements. Noncompliance should not be confused with an approved deviation or exception to the regulations. The Institutional Animal Care and Use Committee (IACUC) may approve exceptions to the regulations and standards if it is justified for scientific reasons by the investigator. An example of such an exception would be the withholding of analgesics or tranquilizers following a surgical procedure because they may interfere with test results. It is the responsibility of the principal investigator to provide the appropriate scientific justification in the animal study proposal for withholding pain relieving medication. The IACUC, however, is responsible for evaluating the proposal, and to approve, require modification, or reject the proposal. Likewise, the attending veterinarian also may approve exceptions to the regulations to maintain the health or well-being of an animal. For example, veterinary

TABLE 5.1

Comparison of Animal Welfare Act (1976), 1985 Regulations, and the NIH Guide.

Animal Welfare Act	9 CFR Previous Regulations and Standards	New Regulations and Standards	PHS Policy/NIH "Guide"
1. The handling, housing, feeding, watering, sanitation, ventilation, shelter from extremes of weather and temperatures, adequate veterinary care, and separation by species.	PART3—STANDARDS, to include Subpart A-F	Part 3—Standards, A+D more specific guidelines on housing, ventilation, shelter, sanitation, for dogs, cats + NHPs.	The *Guide* contains comparable provisions, in some cases general and in other cases more specific (chapters 2,3,4).
2. Exercise of dogs (and) 3. Psychological well-being of primates	Not addressed. Not addressed.	Section 3.7 Exercise & Socialization for dogs. 3.81 Environmental enhancement to promote psychological well-being.	The *Guide* has recommendations on both exercising of dogs and psychological well-being of primates. This is addressed under social environment and space recommendation activities (pp. 12–17)
4. To minimize animal pain and distress in research	Not addressed.	Section 2.31—IACUC Institutional Animal Care and Use Committee (IACUC).	The IACUC must determine that the research project conforms with the institution's assurances and meets requirements. Grant applications must describe procedures to minimize pain and distress (PHS pp. 7,9).

5. For Principal Investigator (P.I.) to consider alternatives to painful procedures	Not addressed.	Sections 2.31—IACUC A responsibility of the P.I., as per the AWA. The IACUC must assure this.	Responsible institutional official shall ensure that other methods are considered (PHS, p. 27).
6. Veterinary consultation in painful procedures.	Not addressed	Sections 2.31 IACUC + 2.33 Attending veterinarian must be consulted by P.I. IACUC assurance & AV duties.	Not specifically addressed. Providing guidance to the users is part of the veterinary care program (Guide, p. 33).
7. The use of tranquilizers, analgesics, and anesthetics.	Addressed in sections 3.10, 3.34, 3.59, 3.84, 3.111, and 3.134—At the discretion of the attending veterinarian.	Section 2.31 (d) (1) (iv) Painful procedures require the use of appropriate sedatives analgesics, or anesthetics unless withholding is justified by P.I.	IACUC must determine that the research project conforms with the institution's assurances and requirements. Grant applications must describe procedures designed to assure that analgesic, anesthetic, and tranquilizing drugs will be used where indicated (PHS, pp. 7–9).
8. Presurgical and postsurgical care.	Not specifically addressed.	Section 2.31 (d) (1) (ix) A function of IACUC to ensure that it is included in animal study proposal 2.33 (b) (5) program of vet care.	The Guide specifically addresses surgery and postsurgical care (pp. 37–39).

(continued)

TABLE 5.1 (continued)

Animal Welfare Act	9 CFR Previous Regulations and Standards	New Regulations and Standards	PHS Policy/NIH "Guide"
9. Against the use of paralytics without anesthesia.	Not Specifically addressed.	Section 2.31 (d) (1) (c) Prohibits paralytics without anesthesia.	The institution official shall ensure that painful procedures should not be performed on unanesthatized animals paralyzed by chemical agents (PHS, p.27).
10. Withholding or pain-relieving drugs or euthanasia only for necessary periods.	Not specifically addressed.	Section 2.31 (d) (1) (iv) (A)—As per the AWA, when approved by the IACUC.	Exceptions should be made by decision of appropriate review group such as IACUC. No exceptions for teaching or demonstration (PHS, p. 28).
11. No animal to be used in more than one major surgery except for scientific necessity or when determined by the Secretary.	Not specifically addressed.	Sections 2.31 (d) (1) (x)—As per the AWA—When approved by the IACUC or special circumstances as determined by the Administrator.	*Guide* discourages; allowed only with committee approval (p. 9). PHS policy states that animals that would otherwise suffer severe or chronic pain or distress that cannot be relieved should be painlessly killed at the end or during the procedure (PHS, p. 27).
12. Exception to standards are allowed only when specified by research protocol. Such exceptions shall be put in a report and filed with the IACUC.	Not specifically addressed.	Sections 2.31 (c) (3)—Language similar to the AWA. Also, explanations for exceptions to the standards are submitted in writing and attached to the annual report. 2.36 (b)(3).	Exceptions should be made by decision of the appropriate review group such as IACUC. No exceptions for teaching or demonstration (PHS, p. 28).

U.S. DEPARTMENT OF AGRICULTURE — ANIMAL AND PLANT HEALTH INSPECTION SERVICE

ANIMAL CARE INSPECTION REPORT

☐ Routine ☐ Reinspection ☐ Pre-license ☐ Attempted ☐ Other

1. LICENSE NO. OR REGISTRATION NO.	2. PAGE 1 OF __
3. DATE OF INSPECTION	4. TIME
5. DATE OF LAST INSPECTION	6. TIME

7. NAME AND MAILING ADDRESS OF LICENSEE OR REGISTRANT

8. ADDRESS OF PREMISES AT TIME OF INSPECTION (if different than Item 7)

9. NO. OF ANIMALS INSPECTED

STANDARDS AND REGULATIONS

"X" if in compliance; CIRCLE Non-compliant items (explain on APHIS FORM 7100, Continuation Sheet); NA if not applicable; NS if not seen.

Group	Item	Dogs (A)	Cats (B)	Guinea Pigs	Hamsters	Rabbits (C)	Primates (D)	Marine Mammals (E)	Mammals (F)	Other
GENERAL	10. Structure and Construction	3.1	3.1	3.25	3.25	3.50	3.75	3.101	3.125	
GENERAL	11. Condition and Site	3.1	3.1				3.75	3.101	3.125	
GENERAL	12. Surfaces & Cleaning	3.1	3.1				3.75	3.101		
GENERAL	13. Utilities/Washrooms/Storage	3.1	3.1	3.25	3.25	3.50	3.75	3.101	3.125	
GENERAL	14. Drainage and Waste Disposal	3.1	3.1	3.25	3.25	3.50	3.75	3.101	3.125	
INDOOR	15. Temperature/Ventilation/Lighting	3.2	3.2	3.26	3.26	3.51	3.76	3.102	3.126	
INDOOR	16. Interior Surfaces	3.2	3.2	3.26	3.26	3.51		3.101		
INDOOR	17. Drainage							3.101	3.126	
SHELTERED	18. Temperature/Ventilation/Lighting	3.3	3.3				3.77			
SHELTERED	19. Shelter from elements	3.3	3.3				3.77			
SHELTERED	20. Surfaces	3.3	3.3							
SHELTERED	21. Capacity/Perimeter fence/Barrier						3.77			
OUTDOOR	22. Restrictions or Acclimation	3.4	3.4	3.27	3.27		3.78	3.103		
OUTDOOR	23. Shelter from elements	3.4	3.4	3.27		3.52	3.78	3.103	3.127	
OUTDOOR	24. Drainage			3.27		3.52			3.127	
OUTDOOR	25. Construction	3.4	3.4	3.27			3.78	3.101		
OUTDOOR	26. Capacity/Perimeter fence/Barrier						3.78	3.101	3.125	
MOBILE	27. Temperature/Ventilation/Lighting	3.5	3.5				3.79			
MOBILE	28. Public Barrier						3.79			
PRIMARY ENCLOSURE	29. General Requirements	3.6	3.6	3.28	3.28	3.53	3.80	3.104	3.125	
PRIMARY ENCLOSURE	30. Space & Additional Requirements	3.6	3.6	3.28	3.28	3.53	3.80	3.104	3.128	
PRIMARY ENCLOSURE	31. Protection from Predators	3.6	3.6	3.25	3.25	3.52	3.80	3.101	3.125	
ANIMAL HEALTH AND HUSBANDRY	32. Exercise and Socialization	3.8	3.8							
ANIMAL HEALTH AND HUSBANDRY	33. Environment Enhancement						3.81			
ANIMAL HEALTH AND HUSBANDRY	34. Feeding	3.9	3.9	3.29	3.29	3.54	3.82	3.105	3.129	
ANIMAL HEALTH AND HUSBANDRY	35. Watering	3.10	3.10	3.30	3.30	3.55	3.83	3.106	3.130	
ANIMAL HEALTH AND HUSBANDRY	36. Cleaning and Sanitation	3.11	3.11	3.31	3.31	3.56	3.84	3.107	3.131	
ANIMAL HEALTH AND HUSBANDRY	37. Housekeeping and Pest Control	3.11	3.11	3.31	3.31	3.56	3.84	3.107	3.131	
ANIMAL HEALTH AND HUSBANDRY	38. Employees	3.12	3.12	3.32	3.32	3.57	3.85	3.108	3.132	
ANIMAL HEALTH AND HUSBANDRY	39. Social Grouping and Separation	3.7	3.7	3.33	3.33	3.58		3.109	3.133	
TRANSPORTATION	40. Primary Enclosure	3.14	3.14	3.36	3.36	3.61	3.87	3.113	3.137	
TRANSPORTATION	41. Primary Conveyance	3.15	3.15	3.37	3.37	3.62	3.88	3.114	3.138	
TRANSPORTATION	42. Food and Water	3.16	3.16	3.38	3.38	3.63	3.89	3.115	3.139	
TRANSPORTATION	43. Care in Transit	3.17	3.17	3.39	3.39	3.64	3.90	3.116	3.140	
TRANSPORTATION	44. Handling during Transportation	3.19	3.19	3.41	3.41	3.66	3.92	3.118	3.142	

45. Identification - 2.38 & 2.50
46. Records & Holding Period - 2.35, 2.75, 2.76, 2.77, & 2.38, 2.101
47. Handling - 2.38, 2.131, 3.111, & 3.135
48. Veterinary Care - 2.33, 2.40, & 3.110
49. IACUC - 2.31
50. Personnel Qualifications - 2.32
51. Other items? YES (If yes, see continuation sheet) NO

52. PREPARED BY (Signature and title)	53. DATE
54. COPY RECEIVED BY (Signature and title)	55. DATE
56. REVIEWED BY (Signature and title)	57. DATE

APHIS FORM 7008 (AUG 91) (Replaces APHIS FORM 7008 (APR 90), which is obsolete.) PART 1 - SECTOR OFFICE
*U.S.GPO:1991-0-526-504/40250

Figure 5.1 Animal Care Inspection Report.

exception could include maintaining an animal in a primary enclosure that does not provide the minimum required amount of floor space because the attending veterinarian may determine that a smaller enclosure is necessary to provide adequate veterinary care for that animal. Exceptions such as these are considered temporary and should be approved only for the necessary period of time.

It is not the intent of the AWA to interfere with scientific investigation. It cannot provide for every conceivable situation or research protocol that may be proposed. Thus, the agency depends on the IACUC to evaluate all animal study proposals. It is inevitable that there will be times when the principal investigator, attending veterinarian, IACUC, and the inspector do not agree on what should be considered an acceptable deviation or exception. When these differences occur, there are methods to ensure an effective resolution. The inspector can refer to the AWA, Regulations, standards, and written policy. If the situation is not addressed, the sector USDA animal care specialist may be called in to help evaluate the situation, and when necessary the animal care staff at the agency's headquarters will evaluate the situation and make a decision. The animal care staff writes and interprets regulations and agency policy, and are knowledgeable in disciplines such as laboratory and exotic animal medicine.

The agency has certain legal responsibilities to Congress to follow up on alleged violations based on noncompliance with the AWA. When noncompliance is documented on an inspection and the violation is flagrant or recurrent, APHIS takes appropriate action by conducting reviews and investigation of alleged violations. There are specific responsibilities on the part of both the agency and the facility when an alleged violation is filed. The institutional official has the legal responsibility to see that the cited deficiency is corrected. The method federal agencies use to adjudicate cases of alleged violation is through an administrative law proceeding. These procedures are published in the Rules of Practice Governing Formal Adjudicatory Administrative Proceedings Instituted by the Secretary (of Agriculture) in the Code of Federal Regulations (CFR) title 7, subtitle A. The following is a summary of the legal steps a facility and the agency can use in resolving differences in interpretation of the regulations:

1. Inspector finds noncompliance on a site visit and gives the facility a specified time to correct the deficiency. Flagrant violations, such as those that jeopardize the health or well-being of an animal, require immediate action. If on reinspection it is determined that either corrective action was taken or a letter of intent was filed with the necessary corrective action outlined within an acceptable time frame, the case may proceed no further or be closed with a letter of warning. Flagrant or recurrent violations generally result in an alleged violation being filed.

2. If the facility does not correct the violation prior to the time specified and reinspection indicates that corrective action was not taken, an alleged violation may be filed against the facility and an investigation begun to gather evidence against the facility.

3. The facility can be issued a letter of warning, or be prosecuted through the Office of General Council where if the facility chooses a settlement, fines of up to $2,000 per violation can be assessed.

4. If the facility chooses to have a hearing, an administrative law judge may render a decision that will have no dollar limit, and may order any specific compliance action.

5. Appeal of the decision by either the agency or the facility can be made to the judicial officer. The resultant judicial officer's decision must be complied with by the agency. However, if the facility wishes it could take the case to the Circuit Court of Appeals.

The majority of cases do not proceed past the first step. For fiscal year 1991, the USDA conducted 3,745 compliance inspections and 242 reinspections at 1,474 registered research facilities with a total of 3,495 sites.

The regulations specify minimal animal care. It must contain specific minimum requirements to act as a legal document. The regulations cannot nor is it intended to cover every possible situation that would be encountered in the animal care community. Interpretations of specific sections of the AWA are distributed through written policy letters and memoranda that help interpret the intent of the Act. These interpretations are a guide and not law; this could lead to some apparent contradictions for inspectors for example, certain types of enclosure environments may fall outside what would be considered the norm of the regulations. These enclosures cause the agency and the facility's manager great concern, as they are not clearly provided for in the regulations, standards for animal care, or Animal Welfare Manual. When this occurs the best method to rectify the situation is to gather all the involved principals, and discuss the specific problems and apparent contradictions in enforcement of the Act. In an overwhelming majority of cases, such a meeting will stop a problem before it has a chance to grow. One should remember, however, that inspectors must enforce the standards as written. It is not their job to interpret the meaning of the Act. The steps previously mentioned is the legal process that allow for this interpretation. This legal process can change an interpretation of the Act simply by settling a dispute and setting a precedent. A facility's manager has the legal right and moral responsibility to explore this avenue of interpretation if they feel that they are providing a better environment for the animals in question.

Aristotle may have been the first to use animals in experimentation, but he has not been the last. If we as scientists wish to see the continuation of animal research for the betterment of both animals and humans, we need to do it appropriately. Compliance with the AWA is not just a legal responsibility but also a moral one.

BENSON E. GINSBURG

6

The Applicability of Regulations Governing Research Facilities for Behavioral Research in Naturalistic Settings

In an effort to insure humane treatment for animals used in scientific research, scientific as well as legislative bodies have formulated standards and imposed methods of compliance (see Bayne and Hendrickson, chap. 4; and Annelli and Mandrell, chap. 5, this volume). These have become more inclusive and more restrictive with each revision, in part to accommodate new advances, but also, and not unimportantly, in response to various self-anointed advocacy groups, some of which have as their objective the cessation of all animal research. To address the legitimate concerns of those who seek improved facilities and conditions has been a constructive, ongoing activity of long standing. To attempt to placate those whose objective is to do away with animal research altogether is a mistaken and futile exercise.

The mission of the scientific endeavor is to advance the frontiers of knowledge for the benefit of humans and other animals alike. The experience of the past bears impressive witness to the fact that this cannot be accomplished without the use of animals. While the medical model is persuasive enough, and it is medical research that is usually the focus of concern, the preservation of wildlife and the maintenance of the ecological balance that binds all living things together also require new knowledge. Studies of animal behavior constitute a link between both sets of concerns. Unfortunately, the standards applicable to medical research are often inapplicable to animal behavior research, which may require a variety of facilities, each suited to the species and the problem being investigated. Present and proposed regulations fail to take this into account.

CURRENT REGULATIONS

The most relevant federal regulations are embodied in the Animal Welfare Act of 1966 and the Good Laboratory Practice Act of 1978, together with various more recent and proposed amendments (Federal Register, part IV, August 31, 1989, and July 31, 1991). The major guideline for users is the NIH *Guide for the Care and Use of Laboratory Animals*. The Animal Welfare Act applies to dogs, cats, nonhuman primates, rabbits, guinea pigs, hamsters, and marine mammals (although legislation is now being proposed to extend this to mice, rats, and birds). Its regulations, as embodied in the original act and subsequent amendments, are monitored and enforced by the Animal and Plant Health Inspection Service (APHIS) of the United States Department of Agriculture (USDA). Among other requirements, research facilities must be registered and must agree to comply with USDA standards. Compliance is enforced by monetary fines, as well as by loss of accreditation, which could result in the shutting down of a project, a facility, or an entire organization.

Research facilities must furnish annual reports that comply with the details of the regulations. Reports include the numbers and types of animals used, the procedures involved, the measures taken to avoid or relieve pain, and the justification of procedures that do not conform to the USDA standards. All animals must be identified as to origin and disposition. Health certificates are required for dogs, cats, and nonhuman primates. The USDA inspects all facilities without prior notice. These inspections can include the examination of records as well as the facilities themselves. The regulations further require certification by a veterinarian that appropriate anesthesia or analgesia is being used, and that euthanasia is humanely and properly performed where this occurs. Licensing of dealers is also provided for, and there are standards for transportation of animals from dealers or from one facility to another.

The Good Laboratory Practice Act applies, in large part, to laboratory studies that have as their objective the development of products that need FDA approval. Again, detailed requirements for compliance are spelled out.

The NIH *Guide* was developed by the National Academy of Science's Institute for Laboratory Animal Resources (ILAR). This has now gone through five revisions and is currently entitled *Guide for the Care and Use of Laboratory Animals.* The *Guide,* as its name implies, embodies a set of recommendations meant to be adapted by investigators to fit their own situations. The *Guide* applies to all vertebrates. Individual grantees and institutions receiving NIH support for animal research are expected to follow the recommendations in the *Guide,* and either to be accredited by AAALAC, which uses the *Guide* as a basis for accreditation, or to maintain an institutional committee on animal care, consisting of at least five members, including a representative of the public and a veterinarian. NIH grantees using animals must submit their protocols to the animal care committee and have approval in order to apply for a grant. Lack of compliance can result in termination of support of a specific project or even in the termination of support for all projects using animals at that institution. This has actually occurred in a number of cases (e.g., City of Hope Research Institute). A comprehensive discussion of current regulations and proposed amendments is to be found in the chapter by George W. Erving III (1985) and in part IV of the Federal Register (August 31, 1989, and July 31, 1991).

BEHAVIORAL RESEARCH: SPECIAL NEEDS

While most of these provisions, if applied with common sense, would serve the biomedical experimenter, the detailed requirements for caging and other aspects of animal housing would not apply in many instances to the type of facility needed for animal behavior research. The standards in this area must be flexible enough to accommodate the range of species and problems under investigation. They could, in some cases, conform to those that have been developed by some of the better zoos, where such research is being done, but additional flexibility is required to permit a much broader range of studies to be carried out. A study of existing facilities, including zoos, deemed appropriate for such investigations would be helpful, and a supplement to the NIH *Guide* that could also be incorporated as an amendment to the Animal Welfare Act should eventually be issued in a form which, like the NIH *Guide*, allows the investigator and the institution appropriate leeway to adjust the facilities to the requirements of the research.

I have been involved with the study of animal behavior since my days as an undergraduate, and published my first study while a student (Ginsburg and Allee, 1942). External support was modest. Even as faculty, we and our students were our own caretakers and, except in serious cases, our own veterinarians. We learned via an apprenticeship system and, of necessity, became adept at cage repair and the construction of pens and fences. Even so, facilities were reasonably standardized, at least to the extent that this was thought to be necessary in order that studies done in different laboratories should be comparable. Although there was good communication among investigators, we had no animal care committees, I was unaware of any strictures governing our facilities and procedures other than previous practice, common sense, and a concern for the health and well-being of the animals. This concern derives not only from ethical considerations, but also from practical ones. Animals that are not in good health are not reliable subjects for scientific investigations. Neither are they desirable as breeders. Valuable breeding colonies have been lost when disease, stress, and other disrupting factors have not been adequately controlled. The scientist, out of self-interest if nothing else, shares the concerns of those sectors of the public who, out of humanitarian considerations, seek to ensure the humane treatment of animals, whether these are in laboratories, on farms, in zoos, kept as pets, or in nature.

As I look back on it, our facilities were at least fair by present standards, and our animals (including mice, guinea pigs, rats, rabbits, chickens, opossums, cats, dogs, ground squirrels, and monkeys) were generally healthy, kept in sanitary conditions, and humanely treated. Then, as now, there were exceptions, and in instances where facilities were substandard, this was usually the result of economics rather than neglect. We were subject to external review from granting agencies and to criticism in the media, sometimes warranted and sometimes not, and usually instigated by what were then called "antivivisectionists."

My first exposure to this type of interaction between the research community and organizations more concerned about animals than about science involved a public confrontation between A.J. Carlson, professor of physiology at the University of Chicago, an eloquent, informed, and effective speaker on the value of animal research, and Irene Castle McLaughlin, who had a flair for the dramatic and a public following as a performer, as well as an antivivisectionist. It was a confrontation of tactics as well as substance. Professor Carlson was presenting data on life-saving procedures that could not have been developed without animal experimentation. Just as he came to his major points, Mrs. McLaughlin made a spectacular entrance, dramatically dressed and leading two Russian wolfhounds (now Borzoi) on a leash. At that point, Professor Carlson no longer had his audience—for a moment. It was only a moment before the lights were dimmed and he began to show slides. Such were the benign confrontations of the 1940s.

Animal Rights: The Tarnished Image

The confrontations of today involve escalated tactics, and are not limited to targeting animal experimentation in the medical area, but also include behavioral research. There is a serious shortage of dollars for upgrading facilities, particularly in this area. The animal welfare regulations, while appropriate for many purposes, do not provide the flexibility needed for the conduct of many types of behavioral research and have not been coupled with increased funding at the federal level needed to facilitate compliance. Research institutions and government agencies remain under increasing pressure from animal rights groups. Their literature conveys the impression that most research involves the use of primates, dogs, and cats. In actual fact, of the 17 million animals used in research in this country during 1987, 85 percent were rats and mice. Of the remainder, there were 61,392 primates, 50,142 cats, and 180,169 dogs. By contrast, 12 million dogs were killed in animal shelters in the same year (*Newsweek*, December 16, 1988, pp. 50-59).

Nevertheless, research with animals is a growing cause célèbre among the 7,000 or more animal protection groups with their estimated 10 million members and more than $50 million budget. They have been effective in blocking or delaying the construction of new facilities and in lobbying for more restrictive procedures and guidelines, while opposing research funding. The concern of the behavioral scientist must be with reemphasizing the importance of animal research in general, behavioral research in particular, and achieving flexible guidelines and funding to permit the often unconventional facilities necessary for behavioral research to be constructed and approved. We have been remiss in not helping to bring about a realization of the importance of behavioral research, particularly to legislators and the lay public, including an appreciation of the facilities needed for the proper conduct of such research and the standards applicable to animals used in behavioral studies.

SOCIETY'S RIGHTS TO NEW KNOWLEDGE

There are many types of behavioral research with animals. These include the effects of brain lesions, the use of operant conditioning, neuro-endocrine studies, studies of selective breeding for behavioral endpoints, studies of the effects of various conditions of rearing, of maternal behavior, agonistic behavior, communication, dominance hierarchies, and social organization. Behavioral studies, therefore, fall into different domains, ranging from those identical or similar to other areas of biomedical research, such as determining the behavioral effects of various drugs, to those that attempt to simulate natural conditions in a controlled fashion. My concern is with all of these, but in particular with the latter.

The concern of the scientific community, which should be the concern of society in general, is that high quality research for purposes of advancing fundamental knowledge that serves to benefit animals as well as humans should be encouraged and facilitated. Knowledge concerning the behavior of animals, including nondomestic animals, under natural conditions has too often been thought of as merely satisfying the investigator's curiosity. It is seen as having lesser practical value than the use of animals for biomedical experiments and has, therefore, been denigrated. Moreover, there has been a movement within segments of the scientific community that fosters a pejorative outlook on captive studies and promotes the view that the only valid observations are those made in the natural environment of the free-livlng animal.

IMPLICATIONS OF BEHAVIORAL RESEARCH

With respect to the first premise, research on the natural behaviors and capacities of animals, domestic and wild, have, since Darwin, had a significant influence on the interpretation of human behavior based on evolutionary affinitics (Darwin, 1872; Hinde, 1974). Studies of animal ethology were followed by studies of human ethology, in which attempts have been made for direct carry-overs from observations on animal behavior to interpretations of human behavior (Eibl-Eibesfeldt, 1979). Humans carry with them as an evolutionary legacy the physiological mechanisms and the associated behaviors that have evolved in other mammalian, and particularly those primate, species with which we share the majority of our genes (King and Wilson, 1975). The field was thought to be of sufficient importance to warrant its recognition in the form of a Nobel Prize. More recently, sociobiology has preempted the role of applying an evolutionary interpretation to human behavior, where, in common with other animals, the natural or biological propensities are seen as serving to maximize the chance of perpetuating one's genotype (Dawkins, 1976; Trivers, 1971). Thus, social behavior and various reproductive strategies have been analyzed from this point of view (see also Buck and Ginsburg, 1991).

The impact of such animal behavior studies in interpreting our own behaviors has been profound. Speculations regarding the nature of human nature, including such questions as whether or not we are a violent species, and whether the tendency for destructive behavior (including war) is in our genes, have been informed by the concepts and generalizations deriving from animal studies, and have found their way into the social sciences (Ramirez et al., 1987). Clearly, the importance of being able to do good research in this area, and particularly of testing the validity of these hypotheses and the limits of their applications, is essential and important for the understanding of the bases of social behavior and social institutions. Otherwise, we are left with global hypotheses that simply represent points of view and have little possibility of being tested. Similarly, information derived from isolate-reared monkeys (Harlow and Harlow, 1962), studies of social hierarchies, and other now classic investigations in the field of animal behavior must continue to be conducted under adequate conditions in order to enable us to generalize from these findings in a valid manner.

THE ROLE OF THE LABORATORY IN BEHAVIOR RESEARCH

Field observations by themselves are often fragmentary, cross-sectional, and situation-dependent. In order to validate our conclusions from these, investigations simulating natural conditions, but carried out in a controlled fashion, are necessary (Ginsburg, 1987). At present, when we are faced with massive encroachment upon natural habitats all over the world, leading to the endangerment of many species, ecological and behavioral studies become essential for the preservation of such species and the possibility of building up and maintaining back-up gene pools in appropriate captive facilities.

Observational studies in the wild, as well as in captivity, are situation-dependent. The classic studies of Sir Solly Zuckerman consisting of zoo observations left the impression that the major behavioral activities of his primate subjects were fighting and reproduction (Zuckerman, 1932). Comparison with field and with later captive observations indicated that this was an artifact of a restricted environment, and that behavior in other environments, and particularly in the field, resulted in a completely different view of normative behavior. Zuckerman's observations and interpretations were not wrong, but situation-dependent.

The now classic Harlow studies of monkeys reared with artificial surrogates (Harlow and Harlow, 1962) and isolation studies performed with other social species, including studies of deprived human children (Bowlby, 1951), led to the conclusion that irreparable damage was done, and that normal social behavior was no longer possible. Later studies by a number of investigators, who provided "therapy" for the isolate monkeys using younger monkeys as behavioral pacemakers, demonstrated that varying degrees of recovery were possible (Suomi and Harlow, 1972). Carter (1988), working with aberrantly reared chimps in a natural setting, was able to bring them in a state of normal social behavior. Similar "therapy" with abandoned, isolated, or maltreated children who, in addition to their other deficits, had not learned to communicate by speech, have also been successful, even when the deprivation extended to ages where recovery was thought to be impossible (Koluchova, 1972, 1976). These owe much to the animal investigations. The American Psychoanalytic Association, for example, has been sponsoring panels at their annual meetings over a period of years focusing on the relevance of animal behavior research for understanding aspects of human behavior and therapy.

In our own studies of wolves isolated at early ages, including from birth and kept in isolation for up to 10 months, two distinct modes of restoring normal social behavior were developed (MacDonald and Ginsburg, 1981). Each of these modes used a different principle from the other, and both used different approaches than those found to be successful in isolate monkeys.

In our work with the socialization of adult wild wolves, the size and design of the runs had to take into account the flight distance of the animals (Woolpy and Ginsburg, 1967). Quite different conditions were needed to carry out the logistics for the isolation experiments mentioned above, and still others for the effects of social stress, which, for some paradigms, included crowding. For outdoor observations in population fields, it was obviously impossible to exclude all vermin, yet we were cited for failure to do so. Difficulties also arose over our isolation studies, during which no inspections could be mounted without disturbing or in other ways affecting the isolates. Persons who would not think of lifting the cover of a Petri dish and thereby contaminating the cultures, fail to see the analogy when it comes to interfering with the necessary conditions of a behavior experiment in which their intrusion would constitute a similar contamination.

Even field observations are condition-dependent. The prevailing views of chimpanzee behavior have recently been modified by Jane Goodall's observations in Gombe of territorial conflicts between two groups amounting to virtual warfare (Goodall, 1986). The interpretation of behaviors seen in the field has been explicated by observations of captive groups under controlled conditions. Such studies have helped us understand many other aspects of "natural" behaviors, including the development and evolution of social bonding and social structure as these occur in animal societies and as the principles derived from animal studies are applied to humans. Two additional examples will suffice.

One has to do with observations of captive wolf packs reared in a large population field. Where the field is surrounded by an opaque barrier and the animals are unhandled, they retain the wariness characteristic of the wild, in contrast to animals in identical enclosures, where there has been no visual barrier (Ginsburg, and Schotte, 1979).

An example of attempts of direct carry-over from animal observations to humans is that of the interpretation of factors involved in mother-infant bonding. In a recent discussion and evaluation of the biological and ethical considerations involved in human surrogate motherhood (the Baby M case), an argument was put forward, based, in part, on rhesus monkey studies, that significant aspects of bonding occurred while the fetus was in utero, and certainly after birth, that would make it psychologically and biologically difficult for a woman who had borne a child not to be traumatized if she had to give it up, even if she never saw her baby (Bard and Kurlantzick, 1990; Fox, 1988).

The importance of having reliable data, including the spectrum of variability characteristic of the species and ways in which an individual might be characterized in terms of predicting behavior, is extremely important. It is necessary, therefore, to be able to make controlled observations and to manipulate conditions in order to arrive at reliable interpretations of animal behavior that will permit one to critically test hypotheses and the limitations of their applications. The manipulations must be appropriate to the scientific questions that need to be answered, and there is no possible substitute that could replace the animals themselves as subjects and objects of such studies.

ANIMAL RIGHTS AND HUMAN PRIVILEGES

Ethical considerations and the objectives of the experimenters converge in the concern that the animals should be maintained in good health and under physical conditions appropriate for that species. However, no procrustean bed embodied in a set of regulations is applicable to the various situations that need to be investigated. While it is necessary for facilities to be supervised, no set of proscriptive guidelines will be applicable to the range of situations and species that will continue to be the subjects of behavioral investigations.

ETHICS

Animal research has been opposed on ethical grounds. Singer (1975), for example, raised the question of whether we humans have the right to use animals for our own benefit. Along with other animals, we, as a species, evolved as predators in order to survive. The erratic nature of hunting has been replaced by the more efficient reliance on animals kept, bred, and raised for food. This became a necessity with growing and denser human populations. We also came to understand our own anatomy by virtue of becoming familiar with the viscera of our prey. Our first studies of animal behavior familiarized ourselves with their habits in order to avoid falling prey to them, and to increase our effectiveness as hunters and, later, as farmers, breeders, trainers of horses, dogs, and other animals on which we relied for transportation, hunting, herding, guarding, and other functions. As civilization advanced, animals were used and continue to be used to expand our knowledge of disease and to develop life-saving techniques and therapics. Our need to better understand behavior at the molecular, individual, and societal levels has also been served by observing and experimenting with animals. Ultimately, we share our knowledge with them by applying it to their welfare and to their conservation under ever more threatening conditions. While we could possibly persist as vegetarians and clothe ourselves in various synthetic products, we could not have developed vaccines, surgical techniques, medications, or come to an understanding of our own behavioral capacities, evolution, and ecology without studying ("exploiting") animals. In response to Singer's question, we have no alternative if we are to advance and survive as a species. As a question of ethics, we have an obligation to be humane, an obligation that we do not yet accord to our fellow humans, as a perusal of any newspaper on any day of the week will show (Ginsburg, 1979). As a question of right, we are compelled by the biological imperative of self and species preservation.

ANIMAL CARE REGULATIONS: A PERSONAL RETROSPECTIVE

As mentioned earlier, I have been continuously involved in studies of animal behavior for many years. Along with most of my colleagues in the scientific community, I have also been concerned about the ethical aspects of what we do and about the health and welfare of our animal subjects. In this context, I served on the Advisory Committee for the Animal Resources Branch of the NIH, which later also became a study section. I also served as a member of the Accreditation Council for the American Association for Accreditation of Laboratory Animal Care (AAALAC) and continued after that to be involved in the monitoring of the guidelines as a site visitor. I have been on the animal care committees of two universities, and served as chair for one of these. I have, therefore, seen and thought about the problems both from the point of view of the investigator and from the point of view of helping to develop and monitor guidelines for appropriate facilities. I have also been concerned with these as the head of a department where behavioral and biomedical research with animals was being carried out. Nor have I escaped the controversies spearheaded by various groups expressing legitimate concerns for the welfare of animals, as well as having to deal with those extremists who see all animal research as cruel and unnecessary. I have spoken with legislators and lobbyists, and have been told that their heaviest mail is from the animal rights movement and that the response from those who see animal research as a social and scientific necessity has been meager and inadequate on two fronts. One is that of input to our senators and representatives, and the other is that of the education of the lay public, many of whom have been led to believe, based on unfortunate actual instances, as well as propagandistic distortion of facts, that animal experimentation is, in fact, cruel, callous, and unnecessary. They have the impression that most animal research is done with dogs, cats, and monkeys, when, in fact, the overwhelming majority of animal subjects are mice and rats (*Newsweek*, December 16, 1988, pp. 50-59). They argue that computer simulations can take the place of animal experimentation when, in fact, the computer can only deal with the information that it is fed, and it is this information that needs to be augmented and further researched. Many are also convinced that experiments are needlessly repetitious, although procedures of peer review make this unlikely.

Efforts continue to be made to further amend the Animal Welfare Act in ways that are unnecessarily restrictive and would result in great and unnecessary expense with no recommendations for provisions to defray these. An example is that of again mandating changes in cage size, often unsupported by hard evidence, that would mean the replacement, at great cost, of cages in most of our animal-holding facilities.

We have a responsibility for educating our legislators and the public to the importance and the requirements of behavioral research, and to develop recommendations and suggestions for flexible and appropriate guidelines and for funding in this area. This volume represents a step in that direction.

III. PHYSICAL DIMENSIONS AND MANAGEMENT CONSIDERATIONS IN THE DESIGN OF NATURALISTIC ENVIRONMENTS

EMERSON L. BESCH
GEORGE V. KOLLIAS, JR.

7

Physical, Chemical, and Behavioral Factors in Large Low-Density Naturalistic Animal Facilities

Animal environments include all internal (e.g., disease organisms, parasites) and external (e.g., air temperature, humidity, illumination, social interactions) nonhereditary conditions under which an animal lives (Hafez, 1968). Whether random-source or purpose-bred, endothermic animals tolerate a wide range of environmental conditions through physiological, morphological, and behavioral adjustments that are genetically determined or environmentally induced (Prosser, 1964). Different environments likely pose different problems but, even in the same habitat, not all animals respond similarly to environmental influences. Survival requires that animals undergo compensatory changes to multiple deviations in their surroundings. Through natural selection, animal species possess homeostatic adaptations that allow survival, assure reproductive performance, and determine the ecological niches or geographic ranges of the populations (Prosser, 1964).

The environmental circumstances that induce changes in the animal are termed stressors. Adaptates are the observable consequences of the resultant adaptation. Thus, adaptates are increments or decrements in whatever is being measured (Adolph, 1956). Although stressor indices such as elevated plasma corticosteroids (Selye, 1936) or behavioral changes (Dantzer and Mormede, 1983) have been suggested for evaluating the appropriateness of the animal management procedures or housing conditions, health, productive, and reproductive traits provide more reliable indicators of the well-being of animals (Sadleir, 1975). Accordingly, characteristics of well-adapted animals include high resistance to disease, longevity, low mortality and morbidity rate, reproductive performance, and normal behavioral patterns (Duncan, 1981; Hafez, 1968).

It is well known that changes in an animal's environment may significantly modify biological response (Lindsey et al., 1978). Nonetheless, many research studies do not provide information regarding the ambient conditions of the experimentation. For example, Lang and Vesell (1976) analyzed 4,080 articles in eight journals and found that only 45.5% reported the sex of the species used, 80% the population density, 6.0% the photoperiod, 6.5% the ambient temperature, 1.2% humidity, and 0.2% the bedding material. It is not clear whether these factors were merely unreported or not considered. What is clear, however, is that all individuals involved with animals used in research, testing, experimentation, or exhibition should be familiar with and sensitive to all factors that will alter the animal's environment. Further, the requirements for the humane care, handling, and housing of animals should receive a high level of attention on a continuing basis.

NATURAL AND ARTIFICIAL ENVIRONMENTS

Guidelines (ILAR, 1985) for facilities and operating standards (CFR, 1989) have been developed for the care and use of laboratory animals used for research, teaching, testing, experimentation, or exhibition purposes. Although the latter excludes animals used or intended for use as food or fiber, zoo animals are included (CFR, 1989). Animal facilities, whether indoor or outdoor, should provide appropriate ambient temperatures, ventilation, lighting, and shelter from sunlight or inclement weather to provide for the health and to prevent discomfort to the animals at all times. Further, the animal environments should be appropriate for the species and its life history. While there have been several reports on the housing and environmental requirements for laboratory animals (ILAR, 1965–80; Besch, 1980, 1985, 1990) and agricultural animals (Anonymous, 1988), little information is available in a single source document regarding the environmental requirements for animals in large, low-density naturalistic settings.

In nature, selection assures that each animal in a population is able to tolerate the demands of the environment, survive, and reproduce offspring. Because of variability within populations, some individuals are able to survive and reproduce under extreme environmental conditions. Although the habitats of animals are related to those of their ancestors, the subsequent genotypes possess the ability to thrive in environments different from their ancestors (Hafez, 1968). Animals typically used in research (e.g., albino rats and mice) were derived from wild ancestors and, through generations of breeding, became adapted to their laboratory environments (Lane-Petter, 1963). While their physical appearance may not have changed appreciably, their dietary requirements, susceptibility to disease, freedom to roam, and behavior all differ between wild and laboratory-bred animals or counterparts. Further, native environments differ from zoo environments in that the latter include such features as housing, diet, cage mates, internal flora (i.e., bacteria) and fauna (e.g., parasites), keepers, and visitors (Ratcliffe, 1968).

Some animals (e.g., rats) moved from their natural habitat possess the adaptability to survive and reproduce in captivity (Hale, 1969). Animals like the cheetah behaviorally adapt to environments of wildlife parks but must be provided secluded areas that allow courtship and successful mating (Eaton, 1971), while others (e.g., cottontail rabbit) display difficulty in surviving in captivity (Hediger, 1964). Captive environments should closely simulate the natural environment in meeting the biological requirements of the animals (Berg, 1987); if they do, there is a high probability that the species will be able to survive and reproduce in captivity (Hediger, 1969). Moreover, the range of species successfully bred and their survivability—not the variety of exhibits—likely determines the quality and reputation of a wildlife park or a zoo (Eaton, 1974; Martin, 1975).

Animals in naturalistic environments typically live in social groups that allow for expression of normative social behavior, organization into species-typical units, and reproductive and parental behavior (McBride, 1976). Therefore, facilities should provide spacious, rich, and natural physical environments for normal behavioral expression of the animals and unobtrusive viewing by the public (Schassburger, 1987) without jeopardizing the health and safety of the animals or the human visitors. Preventive medicine and health care programs should address both the medical needs of the animals and the environmental and social influences related to health (Bush et al., 1987; Gibbons and Stoskopf, 1989).

Natural settings should be designed to avoid using toxic or otherwise inappropriate vegetation, provide good drainage to limit the contamination of the soil or other substrates with potentially pathogenic bacteria, protozoa, and helminths (Fradrich, 1987), furnish both natural and artificial lighting and varying photoperiods to benefit reproductive fertility and health, and provide holding, quarantine, and off-exhibit areas for the treatment of animals outside of public view and for seclusion of stressed animals (Bush et al., 1987). Regarding the latter, it has been reported that the proximity of animals to visitors as well as visitor density contribute to medical problems in bottle-nosed dolphins (Gibbons and Stoskopf, 1989).

Knowledge and understanding of how animals communicate is an important ecological aspect of the behavior of domestic species and necessary for the preservation of wildlife species or management of domestic species (McBride et al., 1967). In addition to chemical signals (pheromones), animals communicate through vocalization, visual cues, and behavior (Whitten and Bronson, 1970). Thus, when developing guidelines for successful confinement, breeding, and management of animals in naturalistic environments, consideration must be given to the biological needs of the animal, the nature of the ancestral environments (Lane-Petter, 1963; Martin, 1975), and the ecology and behavior of the species (Smith, 1974).

It has been suggested that the design of naturalistic animal enclosures should contain but not be limited to several factors. In the order of importance, they include (Durrell, 1976):

1. Needs of the animal.
2. Needs of the staff responsible for the daily care of the animal.

3. The public who visit to view and learn about the animal.
4. Aesthetic aims of those involved in construction and maintenance of the facility.

If research is to be included in the master plan of the zoological garden, input from behavioral scientists, veterinarians, and keepers should be solicited during the facility design planning stages (Schassburger, 1987).

Space Considerations

In captivity, animals are not only deprived of their freedom but also may be exposed to unnatural environments. The effect of the latter is reduced or minimized through use of naturalistic environments that are appropriate to the needs and life history of the species. The quality of an animal's environment should be influenced by the animal's needs rather than by anthropomorphic perceptions (Ross, 1960). A clean feeding trough or nesting box is of less importance to the animal than a branch to climb or sand in which to scratch. Humans are obsessed with cleanliness as evidenced by attempts to provide odorless animal facilities but animals rely on odors as scent marks to delineate territory (Bowen and McTaggart Cowan, 1980; Ralls, 1971).

Animal territories often are described in terms of either the area that belongs to the whole species (biotype) or a social unit (territory). Examples of sizes of territories or home ranges are contained in Table 7.1. Because several individuals of the same species may inhabit the biotype simultaneously, these individuals must share the area by limiting their activities to territories clearly defined by optic, acoustic, or olfactory factors (Hediger, 1964). Nevertheless, territories of many species overlap, predators have bigger territories than herbivores, small animals possess small territories, large animals possess large territories (Leopold, 1986).

Animals taken from the wild to the captive state are exposed to a diminished range of movement, or territorial size. The resultant confinement in the form of a cage, pen, or paddock, often is thought to be harmful to the animal. The rationale for this view is that the animal has limitless amount of space to roam in its "wild state" but only a restricted area in captivity. Actually, the "free-living" animal occupies a territory of limited extent (Hediger, 1964). Within limits, the quality of the environment appears to be far more important than the secondary factors that result from the

TABLE 7.1
Home Range Sizes for Various Animals
in North America

Animal (species)	Mean Range (km²)		Reference
	Male	*Female*	
Arctic fox			
(*Alopex lagopus*)	2.9	2.9	Speller, 1972
Black bear			
(*Ursus americanus*)	42.01	5.0	Garshelis & Pelton, 1980
Bobcat			
(*Felis rufus*)	3.0[a]	1.5[a]	McCord & Cardoza, 1982
	73.0[b]	43.0[b]	McCord & Cardoza, 1982
Coyote			
(*Canis latrans*)	14.2	13.3	Bowen & McTaggart Cowan, 1980
Gray fox			
(*Urocyon cinereoargentus*)	3.2	3.2	Samuel & Nelson, 1982
Gray squirrel			
(*Sciurus carolinensis*)	0.008	0.005	Flyger, 1960
Lynx			
(*Felis lynx*)	19.4	15.5	Saunders, 1963
Mountain sheep			
(*Ovis canadensis*)	3,215	3,215	Welles & Welles, 1961
Mule deer			
(*Odocoileus hemionus*)	12.4	10.6	Rogers et al., 1978
Pocket gopher			
(*Thomomys bottae mewa*)	250–445[c]	120–240[c]	Howard & Childes, 1959

[a]Alabama
[b]California
[c]square meters

diminished space (Hediger, 1964). That is, the lack of free choice of food, inability to choose an optimum micro-climate, diminished capacity to avoid enemies, all are more important considerations than a restriction in the possibility of movement (i.e., limited muscular activity).

Wild animals possess a defense reaction to escape from their enemies (especially humans), whenever there is an encroachment on the flight distance. Unless the animal can elude the enemy and

TABLE 7.2
Flight Distance for Various Species

Species	Flight Distance	Reference
Mammals		
Howler monkey	30–40 ft.[1]	Chapman, 1929
Giraffe	150 yds.[2]	Kearton, 1946
Giraffe	25 yds.[3]	Kearton, 1946
Prong-horned antelope	500 yds.	Heller, 1930
American bison	250–400 yds.	Garretson, 1938
African buffalo	80 yds.	Selous, 1967
Red deer	50–100 yds.[4]	Darling, 1937
Birds		
Ostrich	150 yds.	Kearton, 1946
Herring gull	15–20 yds.[5]	Goethe, 1937
Herring gull	30 yds.[6]	Goethe, 1937
Sarus crane	30–40 yds.	Champion, 1934

[1] In trees
[2] Approached by man walking
[3] Approached by man in motor-car
[4] When being fed
[5] When irritable
[6] Approached by wolfhound

get beyond the flight distance, it will remain in a state of tension or anxiety. Therefore, it would appear that the concept of flight distance should serve as the basis for determining the space that an animal requires for both psychological and physiological reasons. As a general rule, large species have a long flight distance and small species a short one (Hediger, 1955). Utilizing flight distances for various animals (Table 7.2), the minimum size of primary enclosures can be calculated because, in theory, the smallest primary enclosure must be a circle of a diameter twice the flight distance (Hediger, 1964).

Temperature and Humidity

The physical factors of an animal's environment (Figure 7.1) may be described in terms of mass (i.e., gaseous and particulate contaminants and water vapor) or energy (i.e., heat, sound, and light) (Besch, 1980). Because all life exists in an energy environment, there is continuous interaction between animals and their environment. Heat gains may result from solar, reflected, or infrared thermal radiation, while heat losses may result from conduction, convection, radiation, or evaporation (Galineo, 1964). Animal body temperatures are regulated by either environmental heat (ectotherms) or metabolic heat (endotherms).

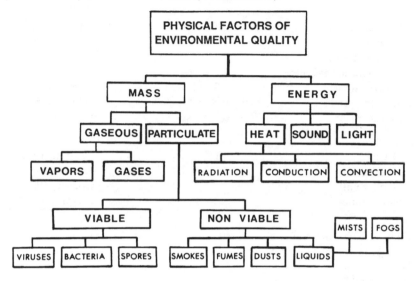

Figure 7.1 Environmental Quality Described in Terms of Physical Factors (from Besch, 1980).

Through acclimatization, endothermic species are able to adjust to their environments either behaviorally or physiologically. Their heat production and heat loss mechanisms maintain body temperature at a constant value over a range of environmental temperatures (Herrington, 1940). This zone of endothermy varies for different species and represents the limits of homeostasis. Within this range is a narrower range (thermoneutral zone) of ambient temperature in which heat gain and heat loss are at a minimum and the animal is able to maintain basal heat production through behavioral adaptation (Stainer et al., 1984). With either declining or elevating temperatures, when the limit of the thermoneutral range is exceeded, metabolic heat production increases; the ambient temperature at which this occurs is called the critical temperature (Herrington, 1940). Although endotherms are able to physiologically adapt to cold more readily than to heat, in general, those adapted to higher temperatures have a higher thermal neutrality (Galineo, 1964).

Metabolic rates of endotherms are influenced by environmental temperature as well as duration at that temperature and, because the thermal environment in which the animal lives is constantly changing, metabolic rates adjust accordingly (Galineo, 1964). It has been reported that in rats, prior thermal history affects the thermoneutral zone and alters the set-point temperature around which thermal responses are regulated (Gwosdow and Besch, 1985). Further, endotherms living in cooler environments have a higher basal metabolic rate than those living in warmer ones (Galineo, 1964).

The effect of changes in temperature and humidity are influenced by the physical state of the animal and whether the animal can physiologically moderate these changes (Weihe, 1976a). As an example, healthy Przewalski horses have been shown (Martin, 1975) to adjust to temperatures of zoo environments (-0.7° to 19.2°C) quite well even though the range differed from the natural state (-2° to +2°C). Reptiles, on the other hand, should be housed under conditions that allow them to attain and vary from a preferred body temperature (Martin, 1975). Because this may be near the upper limit of the thermoneutral zone, burrows, dens, or climbing devices should be provided (Martin, 1975). Other considerations, such as thickness of hair coat, are important because it has been reported that sheep with fleece coats have a significantly different lower critical temperature than do sheared sheep (Table 7.3).

The body temperatures of animals also have been observed to cycle during the day and season. For example, the rectal temperatures of a healthy camel may vary between 34° to 40°C; in the healthy donkey, in summer the range is 36.4° to 38.4°C and in winter 35.1° to 38.1°C (Schmidt-Nielsen et al., 1957). Goats exposed to incrementally increasing temperatures between 20° to 40°C displayed increased water consumption but decreased time in roughage consumption, mastication rate, aggressiveness, time spent standing, and movement. The opposite effects were observed when temperatures were decreased from 20° to 0°C (Appleman and Delouche, 1958).

In naturalistic captive environments, temperatures ideally should be similar to those in the wild. This may be mitigated to some extent by factors that include species, geographic location, and the interactions between animals and humans in these facilities. Temperature and humidity ranges for housing laboratory animals (ILAR, 1965–85) are similar to the thermoneutral zones (TNZ) for the respective species (Besch, 1985). Thermoneutral temperatures, therefore, may be useful in developing ambient temperature guidelines for animals in naturalistic environments. Although protection from extreme changes in environment is provided, animals should be allowed to experience seasonal and climatic changes.

TABLE 7.3
Thermoneutral Zones and Body Temperatures of
Selected Endothermic Animals

Animal	Thermoneutral Temperature, °C		Body T	Reference
	Low	High	(°C)	
Armadillo	28	38	34.5	Galbreath, 1982
Cat	24	27	39	Forster & Ferguson, 1952
Chicken	22	27	41.5[1]	Meltzer et al., 1982
Deer (mule)	-9	9	37.1 to 40.6	Mackie et al., 1982
Dog				
Eskimo dog pup	-30	30	38.6	Scholander et al., 1950
Domestic	23	27	38 to 39	Hammel et al., 1958
Fox (white)	-30	30	38.1	Scholander et al., 1950
Goat				
Mountain	-20	20	—	Krog & Monson, 1954
Sheared	20	27	38.3	Lee et al., 1941
Goose				
Domestic	17	28	41.0[2]	Benedict, 1938
Domestic	18	25	41.0[2]	Brody, 1974
Mountain Sheep (Bighorn)	-20	10	39	Chappel & Hudson, 1978
Nonhuman Primate				
Macaca mulatta	25	31	39	Johnson & Elizondo, 1979
Saimiri sciureus	25	35	39	Stitt & Hardy, 1971
Pig	12	24	38.4	Lee et al., 1941
Pigeon (domestic)	20	30	42.2[2]	Riddle et al., 1934
Pocket gopher	26	32	36.9	Gettinger, 1975
Rabbit	28	32	39.5[3]	Lee, 1939
Raccoon	28	32	38.2[3]	Scholander et al., 1950
Sheep				
Fleece coat	0	27	38.8	Lee et al., 1941
Sheare	20	27	38.8	Lee et al., 1941
Turkey (domestic)	20	28	41.2[2]	Brody, 1974
Weasel	18	25	39[4]	Scholander et al., 1950

[1]From Richards, 1970
[2]From McNab, 1966
[3]From Folk et al., 1957
[4]From Brown and Lasiewski, 1972

Although the facilities and operating standards (CFR, 1989) for housing warm-blooded animals address factors such as ambient temperature, ventilation, and lighting, the subject of humidity is not emphasized and little has been reported on the independent effects of relative humidity on laboratory animals (Besch, 1985). Nonetheless, this is an important consideration because it has been reported that susceptibility to disease is influenced by humidity (Baetjer, 1968). There is evidence to suggest that low humidity is related to an increased incidence of upper respiratory tract infections (Clough and Gamble, 1976) and has been shown to cause ringtail in mice (Nelson, 1960), rats (Njaa et al., 1957) Flynn, 1959), elephant shrews (Hoopes and Montali, 1980), and hamsters (Stuhlman and Wagner, 1971). Because evaporation is the only means of heat loss at high ambient temperatures, thermoregulation may be affected at high humidity levels where evaporative heat loss from the animal is either absent or severely impaired.

Ventilation

Facilities and operating standards (CFR, 1989) and *Guidelines for the Care and Use of Laboratory Animals* (ILAR, 1965–85) all emphasize the need for adequate ventilation of animal facilities by natural or mechanical means to provide for the health and prevent discomfort of the animals at all times. Further, those standards direct that animal facilities be provided with fresh air either by means of windows, doors, vents, fans, or airconditioning systems so as to minimize drafts, odor, and moisture condensation.

The importance of indoor air quality and ventilation air quantity has been known for many years (Yaglou et al., 1936). Subsequent studies, directed toward defining the necessary ventilation that would result in providing "odor-free" environments (Munkelt, 1938, 1948; Runkle, 1964), resulted in the current guidelines for ten to fifteen room air changes per hour for laboratory animal facilities (Besch, 1980). Controlling odors through dilution with outside air may add substantially to the heating and cooling load of the facility (Spielvogel, 1978) as well as expose animals to air that may be of less than adequate quality (Anonymous, 1989).

Effectiveness of ventilation is in large measure influenced by the immediate environment of the animal. For animals housed in rooms, the effective ventilation for the animal is the same as for the primary enclosure (room). When the primary enclosure is a cage or pen, control of the microenvironment is the result of indirect (passive or room-coupled) ventilation (Woods, 1978). Further, ventilation effectiveness or air exchange may be limited by cage design or location within a room (Besch, 1975). Other factors that may be used to describe control criteria for enclosed laboratory animals have been discussed elsewhere (Woods, 1980).

Cage design and husbandry practices also may influence concentrations of gaseous and other volatile contaminants (Besch, 1975; Murakami, 1971). Animal cage ammonia levels between 21 (Serrano, 1971) and 700 ppm (Flynn, 1968) have been reported; carbon dioxide has been shown to increase eightfold to 4,517 ppm (Serrano, 1971). In environments containing low NH_3 concentrations, 80% of inhaled ammonia is released without being absorbed (Silverman et al., 1949). Nonetheless, it also is known that at ammonia levels normally encountered in cages, rats consistently exhibit increased rhinitis, otitis media, tracheitis, and pneumonia that are characteristic of murine respiratory mycoplasmosis (Broderson et al., 1976). Environmental ammonia also has been reported to cause increased susceptibility to respiratory infections (Kling and Quarles, 1974), keratoconjunctivitis (Carnahan, 1958), spontaneous corneal opacities (Van Winkle and Balk, 1986) and pulmonary lesions (Lindsey et al., 1978).

Ventilation rate with odor-free air and generation rate of the substance are the two factors that influence the concentration of a gaseous contaminant (Munkelt, 1948). Increasing room air changes per hour or fractional recirculation of room air has limited effect on the reduction of ammonia concentration at equilibrium (Besch, 1985). On the other hand, inadequate ventilation reduces weight gain in beef and other animals (Hazen and Mangold, 1960) and contributes to tail biting in pigs (Van Putten, 1969).

Illumination

Most animals in their natural habitats are exposed to the effects of sunlight, which has a spectrum of about 300–2000nm and it is from the visible light spectrum (390–750nm) that the primary light effects are derived (Goldstein, 1988). The important factors of illumination are quality (i.e., wavelength [in namometers, nm] or color), photoperiod (i.e., the number of hours of light [L] versus the number of hours of dark [D] in a 24-hr solar day), and intensity (i.e., expressed as lux or lumens/m^2). Much has been reported in the literature on photoperiod effects; less is known about light intensity and quality effects (Bellhorn, 1980).

Photoperiodicity has been reported to regulate circadian (i.e., about 24-hr in length) cyclicity in animals (Hasting and Menaker, 1976) although other factors such as temperature also reinforce the cyclic activity. For example, in some environments, high humidity and cold are associated with periods of darkness, while low humidity and heat are associated with periods of light. Some biological rhythms may be seasonal, during which long days are associated with heat while short days are associated with cold. In some mammals and birds photoperiod influences reproductive cycles, migratory patterns, and seasonal changes in feather or hair coat (Folk, 1974). Critical day length (i.e., 10–14 hr L) appears to be the decisive factor in periodic reactions (Bünning, 1967).

Federal regulations for animal facilities contain provisions for natural or full spectrum artificial light (CFR, 1989). Natural light can be provided to enclosures through use of windows or skylights. On the other hand, artificial lighting also can provide some of the desired benefits because both incandescent and fluorescent lighting have qualitatively similar spectral wavelengths compared to sunlight (Bellhorn, 1980). While incandescent light places greater emphasis on red than violet wavelengths, some fluorescent lighting more closely resembles sunlight because of the emphasis on violet and ultraviolet wavelengths (Reiling, 1989).

Although little research has been completed regarding the comparative effects of light intensity on animals, there is concern about what constitutes a safe light intensity in enclosed facilities where artificial light is the only source of illumination. This concern is partly due to the reports of light-induced degeneration of photoreceptor cells in the retinas of rats exposed to 4 days of continuous light at 194 lux (O'Steen, 1970) and that an intensity of 64 lux causes retinal damage in the albino rat (Anderson et al., 1972). Animals in enclosed spaces should be protected from excessive light (CFR, 1989).

Because of the evidence that too much sunlight can be harmful to animals (Sharon et al., 1971; Duke-Elder and MacFaul, 1972), shaded areas should be provided to protect them from exposure to direct sunlight. Also, although it has been reported that high levels of ultraviolet light can produce cataracts in certain laboratory animals (Zigman et al., 1973), the spectral wavelengths from bulbs used in artificial lighting do not appear to pose a hazard to the eyes of humans or laboratory animals (Bellhorn, 1980).

Population Density, Crowding, and Isolation

The chief preoccupation of wild animals is to find safety from predators, including humans. All other factors (e.g., food, water, reproduction) are secondary (Hediger, 1964). In sensing the environment, animals respond in terms of maintenance of homeostasis; the first responses are behavioral and the second physiological (Davis, 1978). The resultant physiological adaptations allow the animal to live in its natural habitat with minimal exposure to environmental stressors. In captive environments, the goal of management is to provide physicochemical and social environments that the animal will not attempt to avoid (Davis, 1978; Hediger, 1964).

Artificial settings differ from natural in that animals may be exposed to social environments that include members of the same species, members of other species, and humans all in the same room or space. Further, captive animals, faced with potential stressors not necessarily encountered in their natural state, may exhibit changes in growth, reproduction, or resistance to disease (Snyder, 1975). Because species differ in their responsiveness to grouping, social species should be housed in groups and others should be solitary (Eisenberg, 1967). Nonetheless, the relations with others of the same species may result in fighting during establishment of social organization (Davis, 1978). This is particularly important when members of the same species are housed together; placing animals of a different species in the same room does not appear to have any social consequences (Davis, 1978).

Social environments include factors such as crowding, population density, and isolation, all of which influence the physiology and affectivity of the animal (Rohles, 1978). Density is an important but different consideration from crowding; the former refers to the number of animals per unit of space while the latter is a perceived lack of space (Stokols, 1972). For example, in studies on the physiological effects of social stress in wild rats, mortality was lowest in control males and in all male groups and highest in male-female colonies (Barnett, 1958). Also, wild male house mice, in confined populations of fixed size, display increases in adrenal weights; size of thymus, seminal vesicles, and testes are inversely related to population size (Christian, 1955).

Density and space also appear to affect the sociality and health of animals. Norway rats develop abnormal patterns of behavior when population is increased in a confined space (Calhoun, 1963). In experiments with rabbits, both sexes respond to crowding in ways that are similar to other species: loss in body weight, loss in weight of organs involved in metabolic function, impairment of reproductive efficiency, and change in adrenal morphology (Myers et al., 1971). Also observed were increased sexual and aggressive behavior and decreased reproductive capacity in female rabbits provided the smallest living space, even though this was accompanied by a decrease in numbers of animals per group (Myers et al., 1971). Although less than that observed when strange monkeys were introduced into the group, aggressive behavior in rhesus monkeys increased when the size of their living area was halved (Southwick, 1969). Grouped rats exposed to procedures such as weighing and injection of saline and several drugs produced striking increases in plasma corticosterone levels compared to isolated rats (Barrett and Stockham, 1965). Trivial handling of rats for 3 minute periods at 2 to 5 days of age results in a prompt increase followed by a rapid lowering in plasma glucose compared to nonhandled rats that show little change over a period of one hour (McIver, 1965).

Animals also respond in other ways to alterations in social environments. Changes in reproductive behavior (Christian, 1955), lactation and litter size (Christian and Lemunyan, 1958), and plasma glucocorticocoids (Johnson and Vanjonack, 1976) all have been reported to be caused by crowding. Further, it has been reported that the pituitary-adrenal responses of pigs exposed to 10 minutes in a new environment were the same as those resulting from exposure to inescapable electric shock (Dantzer and Mormede, 1983) and the increase in plasma glucocorticoids is more pronounced in subordinate than dominant animals (Arnone and Dantzer, 1980).

Chemicals and Drugs

Over the past 35 years, research animals have become healthier, lived longer, and experienced less infectious disease largely as the result of improved laboratory animal environments resulting from use of chemicals, detergents, sanitizing agents, insecticides, and pesticides (Burek and Schwetz, 1980). Although the subject of chemicals and toxins in an animal facility has been covered in detail elsewhere (Lang and Vesell, 1976; Lindsey et al., 1978; Newberne and Fox, 1978), several of the chemicals are of sufficient importance to merit special attention.

In the degradation of toxins and therapeutic agents, hepatic microsomal enzymes (HME) are important but they are particularly sensitive to environmental chemicals and gaseous contaminants found in animal facilities. Sources of these chemicals include insecticides (Conney and Burns, 1972), room deodorizers (Cinti et al., 1976), dietary contaminants (Newberne and Fox, 1978), disinfectants (Jori et al., 1969), and bedding materials (Vesell et al., 1976). Red cedar shavings also have been reported to cause decreased hexobarbital (Vesell, 1967) and pentobarbital (Cunliffe-Beamer et al., 1981) sleep times and increased liver:body weight ratios (Cunliffe-Beamer et al., 1981); vermiculite bedding may cause dehydration (Hastings, 1967). Hepatoma and hepatorenal injury may result from aflatoxin contained in ground corncob bedding (Port and Kaltenbach, 1969) and peanut meal (Newberne and Fox, 1978). It has been reported that immune response may be altered (Wasserman et al., 1969) and lymphocytopenia induced (Keast and Coales, 1967) by insecticides. Phytoestrogens, derived from a soybean product that was a component of the diet, may be one of the major factors contributing to a decline in fertility and in the etiology of liver disease in captive cheetahs (Setchell et al., 1987).

The environmental and genetic factors that influence the response of laboratory animals to drugs have been reviewed elsewhere (Vesell et al., 1976). Although the magnitude of the influence varies with the drug or animal species (Ellis, 1967), ambient temperature also appears to be a major factor (Shemano and Nickerson, 1958). In studies regarding the effects of environmental temperatures between 8°C and 35°C on the acute toxicity of a number of compounds in rats, the vast majority were most toxic at the higher temperatures (Keplinger et al., 1959). There also is evidence to suggest that there is a circadian peak in susceptibility that has a timing similar to other susceptibility rhythms involving the central nervous system (Marte and Halberg, 1961) .

ZOONOSES AND DISEASE PROBLEMS ASSOCIATED WITH ANIMAL ENVIRONMENTS

Laboratory-acquired diseases have been summarized according to the personnel involved (Sulkin and Pike, 1951). In addition to trained scientific personnel, students, animal caretakers, janitors, dishwashers, and other workers were at risk and acquired various types of bacterial, viral, rickettsial, parasitic, or fungal diseases in the laboratory. Diseases such as amebiasis, shigellosis, and tuberculosis affect both humans and animals (Montagna, 1976) and have been reported in zoological gardens (Snyder, 1975). Thus, epizootic and zoonotic diseases are of special importance in naturalistic environments because of concerns for the various species housed there as well as for the zoo personnel and the public.

Zoonotic diseases are commonly found in domestic animals, pets and wildlife. Among these are leptospirosis (Hanson, 1982), yellow fever, West Nile virus, Rift Valley fever (Durojaiye, 1984), and rabies (Behbehani, 1972). A variety of zoonotic diseases, including herpes virus, rabies, mycobacteriosis, *Shigella spp., Salmonella spp., Candida albicans, Entamoeba spp.*, and other protozoal infections have been acquired from pet primates, but laws regulating the importation of nonhuman primates have reduced the hazard of zoonoses (Renquist and Whitney, 1987). Parasitic zoonoses have been identified in islands of the Caribbean Sea and adjoining mainland (Bundy and Steele, 1984). In zoonotic emergencies, the U.S. Department of Agriculture carries out all animal-related disease control and eradication activities in concert with the U.S. Department of Health and Human Services and state public health departments (Atwell, 1987).

There also appears to be a relationship between behavioral stress and disease in animals (Snyder, 1975). For example, mice subjected to the stressors associated with avoidance learning appear to be more susceptible to infection with herpes simplex virus (Rasmussen et al., 1957) or Coxsackie B virus (Johnsson et al., 1959). It also has been reported that male and female albino mice crowded before but maintained individually after infection resisted tuberculosis better than animals housed individually prior to but crowded after infection (Tobach and Bloch, 1956). Resistance to endoparasitism (*Trichinella spiralis*) was decreased in mice that were crowded or that received injections of corticosteroids (Davis and Read, 1958)

Although not disease-related, traumatic injuries also appear to result from behavioral stress. Of all mortality in birds and mammals at the Philadelphia Zoo between 1951 and 1965, 46% was due to injury; the majority inflicted by cagemates (Snyder, 1975). Whether the high incidence of injuries is related to captivity is unclear, but field studies suggest that intraspecific conflicts seldom result in serious injuries in nature; rather, litter mortality generally is density-dependent (Wynne-Edwards, 1962).

Enclosure design, animal grouping, behavioral considerations for each species, and daily husbandry practices all are important factors in the prevention of traumatic injuries of captive wild animals. This has been documented in zoos and other facilities housing captive wild species (Boever, 1972; Wallach, 1970, 1976).

CONCLUSIONS AND FUTURE CONSIDERATIONS

Animal environments have been described in terms of physical, organismic, and reciprocative factors (Rohles, 1978). Those physical factors include humidity, dry-bulb temperature, noise, lighting and air movement; organismic factors include age, sex, genetics, and body type; and reciprocative factors include diet and activity. As independent variables, all these factors are controllable. It therefore is important for researchers and facility managers and designers of low-density naturalistic animal facilities to understand, characterize, and evaluate these factors so that the physiology, affectivity, and behavioral responses of the animal can be interpreted (Stoskopf, 1983).

In defining environmental conditions for animals (regardless of species), physiological and behavioral requirements must be satisfied and stressors that may predispose individuals to disease must be eliminated. This includes obtaining knowledge on which species can be safely housed together. For example, turtles often are a source of *Etamoeba spp.* for snakes (Frank, 1984) and certain nonhuman primates subclinically carry herpes viruses lethal to other primate species (Soave, 1981). Ultraviolet light, properly used, has been reported to be effective in reducing the number of airborne vegetative organisms escaping from animals (Phillips et al., 1957).

Acceptance and utilization of existing information will be helpful in assuring that environments are appropriate for the species and its life history (ILAR, 1985). Where little or no information is available for a particular species, interested persons or groups should identify areas of concern and develop priorities for solving problems and maintaining environmental quality within animal facilities of all types (Besch, 1980).

Preventive maintenance schedules that will assist in minimizing downtime of the lighting, heating, ventilating, and air conditioning of indoor facilities should be developed (Besch, 1980). Further, in addition to observing and caring for animals on a daily basis by qualified personnel, procedures should be established and maintained for providing routine preventive health care and emergency veterinary care on weekends, holidays, and after other regular work periods (ILAR, 1985).

RICHARD J. BRENNER

8

Arthropod Pests: Varieties,
Risks, and Strategies
for Control in
Naturalistic Facilities

Arthropods are the largest phylum of animals on earth (Borror et al., 1976) and their diversity has allowed exploitation of virtually every ecological niche. In their native environs, many biotic and abiotic factors keep populations below pest levels. However, once these constraints have been alleviated by the bountiful nature of artificially controlled indoor environments, populations increase to damaging levels. This chapter will not discuss all of the arthropods that may infest naturalistic facilities. Instead, the primary objective is to provide broad concepts that apply regardless of the pest species. I will define the common denominators of arthropod pests that enable them to thrive in our buildings, why this happens, how it can be prevented, and in instances where it cannot, how it can be managed. Ectoparasites will not be discussed, because problems created by their presence almost certainly will be specific to the host species, its health, and its environment. However, much of this chapter is applicable to constraints on survival of flea larvae (for a review of flea biology, see Dryden, 1989). Further, because cockroaches are among the most pestiferous arthropods in naturalistic facilities, research data on their behavior and ecology will be used to develop concepts for the prevention or resolution of general arthropod problems.

About 4500 species of cockroaches have been classified worldwide. Of these, only about 50 are pestiferous, but virtually every problem species in the U.S.A. was introduced inadvertantly from Africa or Asia (Cornwell, 1968). So, any materials imported into naturalistic research facilities should be inspected thoroughly. Although many species of cockroaches may infest buildings, they can be placed into two general groups: peridomestic and domestic. Peridomestic species are those that live "around the domestic" environment and often in it, but are not obligatorily dependent on buildings. Populations may arise from within structures by introduction of infested goods or equipment, or infestations may result from nearby outdoor habitats. These species are usually large (adults > 3 cm) and many are in the genus *Periplaneta*, including American (*P. americana*), smokybrown (*P. fuliginosa*), Australian (*P. astralasiae*), and brown cockroaches (*P. brunnea*). Other peridomestic species include the Oriental cockroach (*Blatta orientalis*), and the Surinam cockroach (*Pycnocellus surinamensis*). American and Oriental cockroaches commonly are associated with sanitary and storm sewers, and enter buildings through drains and sewer air stacks. Surinam cockroaches commonly infest deciduous compost, mulch, or peat brought into buildings, whereas other peridomestic species enter from nearby outdoor habitats.

In contrast, domestic species are usually small and confined almost exclusively to indoor environments. Common species are the German cockroach (*Blattella germanica*) and the brownbanded cockroach (*Supella longipalpa*). Infestations can result from one or more specimens brought onto the premises in infested materials such as groceries, corrugated board, and infested furniture (Cornwell, 1968; Ebling, 1975).

RISKS FROM ARTHROPOD INFESTATIONS

Aside from regulatory problems with infestations, many arthropods (especially cockroaches) harbor and forage in unsanitary areas, making them suspect in maintaining and disseminating pathogenic microorganisms. A review of literature reporting natural isolations of pathogens of humans found in cockroach populations reveal at least 30 species of bacteria in 16 genera, 4 strains of viruses, 2 species of fungi, 2 protozoans, and eggs of 7 helminths (Brenner et al., 1987). Roth and Willis (1960) compiled an annotated list of cockroach symbiotes that are pathogenic to vertebrates, including some acanthocephalan parasites of primates and various pathogens of canines. Defensive secretions and metabolic wastes also can cause illness if these are contacted or ingested (Roth and Alsop, 1978).

More insidious are the effects of arthropods on allergic disease. Cockroach allergy is now recognized as second only to house dust mites as the most common allergy among asthmatics (Kang and Morgan, 1980). Furthermore, this is a major occupational risk to those who work with arthropods or materials commonly infested or contaminated by them (Wirtz, 1980). Recent studies (Baldo and Panzani, 1988) demonstrate existence of cross-spectrum allergens, suggesting that allergy to one arthropod may result in broad cross-reactivity to many arthropods. The role of arthropod allergen in other animals is not known, but clearly the general health implications warrant every effort to manage arthropod populations safely.

CONSTRAINTS ON ARTHROPOD SURVIVAL

Knowledge of the three primary constraints that limit arthropod survival is essential in understanding how populations flourish in naturalistic research facilities. Arthropods are small and have a high surface-to-volume ratio that makes them prone to rapid loss of water (Chapman, 1971). In general, arthropods captured in the field and placed in a container with water and an acceptable food source will go to the water first. German cockroaches can survive without food for days or weeks, but even in a closed environment, must have liquified water (not just high humidity) every few days (Willis and Lewis, 1957). Air movement further increases water loss (Ramsey, 1935). Thus, the primary constraint on survival is an adequate source of water (Cornwell, 1968) preferably in an environment characterized by high relative humidity and minimal airflow.

The second constraint is harborage, or safe shelter during times of inactivity. To conserve water and energy, and reduce risk of falling prey, harborage is essential. This serves as a base of oper-

ations from which foraging trips for food and water begin. Harborage sites closest to water will be filled first, and the better sites are
dark with physical dimensions allowing both the venter and dor-
sum of the insect to be in contact with surfaces of the harborage
(Mizuno and Tsuji, 1974). Cockroaches also produce an aggregation pheromone that is a component of feces (Ishii and Kuwahara,
1968). This marks suitable harborage sites and ensures that future
infestations will find and utilize this valuable resource.

The final constraint is food. At first inspection, this might
appear paramount, but cockroaches, and many arthropods, are
omnivorous, which enables them to survive on virtually anything
organic. Although they feed heavily on carbohydrates, laboratory
studies measuring weight increase as a function of pure diets
showed greatest weight gain from dried beef steak (Melampy and
Maynard, 1937). In addition, cockroaches have evolved an efficient
mechanism for surviving when their diet is nitrogen-deficient.
Cockroaches have been in existence for nearly 300 million years
(Cornwell, 1968). During that time, a symbiotic relationship has
evolved with a bacteroid, *Blattabacterium cuenoti*, found only in
certain cells (mycetocytes) of cockroaches (Wren et al., 1989).
Researchers have hypothesized that these bacteroids convert
excess nitrogen to uric acid crystals, which then are sequestered
within the fat body. Should the diet become nitrogen-deficient, the
bacteroids convert the uric acid to usable nitrogen (Cochran, 1975;
Cochran and Mullins, 1982). Thus, cockroaches become cannibal-
istic under these conditions, and consume the fat bodies of their
kin. A laboratory colony of German cockroaches, *Blattella germanica* (L), was maintained on no protein for 2.5 years. Although
only a small number of individuals could be sustained, reproduction continued, an extraordinary feat in itself (R. D. Kramer, personal communication). The colony eventually died when the
water supply was exhausted over a long holiday.

CONCEPTS OF PREVENTION AND CONTROL DERIVED
FROM PREDICTABILITY OF ARTHROPOD BEHAVIOR

Focalization Based on Distribution of Resources

Because biotic and abiotic constraints can limit survival of
unwanted arthropods, our objective is to fully discern the ecology
of these potential pests and predict their behavior. The logic is
simple—if we know where they will be, and what governs their

activity, then we can eradicate populations or deny them those factors essential for their survival.

This is best exemplified by describing field research projects conducted on peridomestic cockroaches in Florida (Brenner, 1988). Results and conclusions will show that (1) arthropod behavior is predictable, and is based on probabilities of survival, (2) spatial distributions of pests over time identify principal habitats—by definition, areas where the probability of survival is optimal—and (3) characteristics of principal habitats identify optimal conditions for the species.

The research involved capture-mark-release-recapture techniques on several properties. Ten colored paint pens, representing digits 0–9 were used to give each cockroach a unique 3-digit color code similar to the scheme used by the electronics industry to code resistors (Brenner and Patterson, 1988). Although the smoky-brown cockroach (*Periplaneta fuliginosa*) was the predominant species, concepts are believed valid for others that can infest naturalistic facilities. One objective was to examine the spatial distribution of these species over time to determine whether populations are focused (i.e., concentrated) in easily recognizable habitats. The null hypothesis of uniform cockroach distribution was tested against the alternative of concentrated distribution where conditions are optimum, and that these areas are recognizable to us based on ecological zones. A second objective was to assess the mobility of the species. Systematic sampling was employed, where 100–130 traps were placed at each home in all possible ecological zones (potential microhabitats). Terrestrial zones included grass, mulches, flower gardens, woodpiles, and the base of the houses. Traps were also placed on the trunk of all trees on the property (2 m above the ground), and in some instances at the roof line. Trap coordinates and daily catches were entered into a computer program[1] that develops a visual display of distribution patterns. This mathematically-intensive procedure estimates cumulative number caught, for the duration of the study, at 1-meter intervals throughout the property. Distribution can then be displayed as two-dimensional contour lines or a three-dimensional image showing lines of equal cockroach density.

Figure 8.1 is provided as an example of typical results, and reflects distribution on one acre (0.4 ha) based on data from 103 traps. Even though comparable numbers of tree traps and ground traps were used, analysis clearly showed the importance of trees,

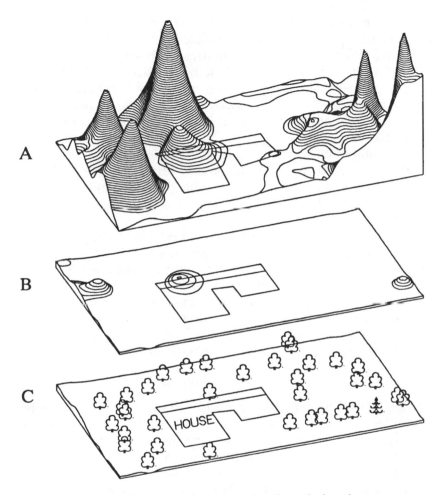

Figure 8.1. Three-dimensional images of cockroach distributions on one acre property based on all traps (A), and ground traps only (B), illustrating the focality and arboreal tendencies of peridomestic cockroaches. C shows location of hardwood trees and one pine relative to the house. Contour lines begin at 1.0 and represent cumulative number trapped in 10 days; intervals are in units of 0.5. See text for further descriptions.

where over 95% of the total number of cockroaches captured were located. The only significant incursion on the house was on the west side in proximity to the tree.

Exactly what was the difference among trees? Why were cockroaches found in only 7 of the 34, trees? Although several factors were examined—such as tree species, diameter, height, and distance to nearest branch—the presence of a tree hole was the only factor that consistently predicted cockroaches.[2] Thus, the null hypothesis was rejected for the alternative; peridomestic cockroaches were concentrated in areas where conditions were optimum, and these areas are recognizable as tree holes. This is the principal habitat or focus where conditions are optimal and the probability of survival is highest. The stable environment can be described as dark, offering protection from predators, providing warmth and humidity, having little air movement, and frequently containing liquid water. Additionally, this biological hothouse, with its attendant microflora and fauna, provides ample food commonly consumed by these omnivorous arthropods (Schal et al., 1984).

Based on studies at several sites over 2 years, the general and specific habitats were discerned (Brenner, 1988). These include, in descending order of importance, (1) tree holes (regardless of species) and the ecological equivalent, such as damp and improperly ventilated attics, voids in walls and voids in concrete block; (2) certain types of leaf litter (there are differences among types [see below]); (3) and wood piles, and their ecological equivalent—that is, wooden decking and other wooden objects with numerous voids.

Strategies of Suppression Based on the "CIA" Concept

Once principal habitats have been identified, the next logical step is suppression of the pests. However, one additional concept is needed to understand how pest populations can be effectively managed. Called the "CIA" (concentrated, immobile, and accessible) concept of insect suppression, it was promulgated originally for mosquito control (Horsfall, 1985), but is valid for virtually any pest. Succinctly stated, long-term pest suppression programs are most likely to be successful if control strategies are directed toward life stages that are concentrated, immobile, and accessible. In the example of the peridomestic cockroaches, the situation is ideal. Concentration in tree holes is augmented by minimal mobility. Even though the maximum distance traveled between successive recaptures was 38 m (for one individual), 72% of all recaptures involved net movement of <1 m (Brenner, 1988). They may move significantly in foraging for food, but they return to the same site, making their behavior predictable. Finally, because these populations are concentrated in easily recognizable regions, the vast majority are accessible.

What is the best strategy for controlling established infestations? Each situation will have unique components, requiring careful consideration as to pest species, locations and extent of infestations, and restrictions dictated by presence of exotic animals in the vicinity, research goals, and the like. However, some general guidelines can be provided. For the larger species, the simplest and least toxic method is trapping. These cockroaches have low reproductive potential, requiring 5 to 12 months to complete development (Cornwell, 1968). Use of baited traps for 2 to 4 weeks may result in virtual eradication for 8 months or more (Brenner, unpublished data). Sticky traps may also be used, but some larger species are capable of escaping. In contrast, reproductive potential of German cockroaches is high (Grothaus et al., 1981; Larter and Chadwick, 1983) and traps are ineffective in controlling populations. However, sticky traps are essential as a surveillance tool to aid in determining focality, degree of infestation, and time of reinfestation (Ballard and Gold, 1982; Reierson and Rust, 1977; Rust and Reierson, 1981; Rust, 1986). Toxic baits, insecticidal sprays with residual activity (especially some synthetic pyrethroids), and new-generation insect growth regulators have been used successfully for this species, although insecticide resistance can be common (Gold, 1985; Cochran, 1989). Choosing the appropriate method depends in part on the risk of nontarget species encountering the pesticides (for general information on control options, see Bennett and Owens, 1986; Mallis, 1982; Ebling, 1975).

PROBLEM-SOLVING IN NATURALISTIC RESEARCH FACILITIES

Facility Design Considerations

Often, specific designs that are necessary for the welfare of the research animal, or to obtain an aesthetically pleasing display, also provide excellent habitats for pests. Problem-solving may require one or many steps, but always the strategy is to reduce the probability that the pests will survive. In the context of the "CIA" concept, if pest species cannot be excluded, at least make them accessible for control programs.

Perhaps the most common problem is with artificial tree holes. This can literally be a tree hole if, for example, a rotted trunk is used as a house for vertebrates. However, the ecological equivalent of treeholes will be more common—voids created dur-

Figure 8.2.   Access point in a gunnite wall constructed from 8" diameter threaded PVC piping. Should arthropod infestations occur within the wall, the cap can be removed for placement of baits or for arthropod-specific growth regulators.

ing the construction process. Gunnite is used commonly in zoos, primate houses, and dens of large carnivores to resemble natural rock formations. In spite of the best workmanship, some cracks will develop, allowing entry of cockroaches and other arthropods to the dark, protected, moist, and warm environment. Access points can be retrofitted or incorporated into the design by placing PVC pipe with a threaded cap through the gunnite (Fig. 8.2). Should the void become infested, the cap can be removed and either toxic baits or traps can be emplaced. Further consideration should be given to whether ventilation to the voids can be increased, so arthropods that may gain entrance will find a less desirable environment. Voids behind wooden, fiberglass, paper mache, or plaster facades also become suitable habitats, especially if a source of water is nearby (Fig. 8.3). If facades cannot be avoided, provide a hinged section to gain access to the recesses behind so that toxic baits, dusts, and the like, can be applied .

Floor drains serve as conduits for American cockroaches that commonly infest municipal sewage systems. Ingress will occur at night, but can be prevented by placing tight-fitting plugs into the

Figure 8.3. Facades create voids that serve as harborage for arthropods and small mammals. If alternative designs that eliminate voids cannot be considered, hinge small sections to allow access for pest control.

drain at night, perforated metal covers, or by using novel hinged cover plates that open downward with water pressure but otherwise remain closed (Scott, 1991). Alternatively, other materials mentioned previously can be used to intercept these invaders (baited traps, sticky traps), enhance desiccation (sorptive dusts [described later]), or kill outright (poison baits).

Other problems occur in the design of landscaping or of laboratory equipment. In many instances, arthropod infestations can be prevented or eradicated simply by eliminating suitable habitats. Like gunnite, stacked rock also creates voids that serve as harborage sites for arthropods and small mammals. Eliminate the habitat by applying mortar or polyethylene foam (expansive) between rocks. Hollow tubular legs of tables, cages, sinks, and benches can become heavily infested with German cockroaches. These habitats can be eliminated by sealing access points with capping, plugging, or filling with foam or caulking compounds. Voids within equipment such as conveyors, refrigerators, autoclaves, and centrifuges provide harborage sites for pests. These often can be managed by removing panels and placing toxic baits inside. Be aware that any room declared "pest-free" may become reinfested quickly if infested portable equipment (carts, cabinets, gurneys) are brought into the room even for short periods Thus, mobile equipment requires special attention.

Sometimes, only slight changes are necessary to make a habitat wholly unacceptable, or to reduce the probabilities of survival. The void behind splash boards of laboratory sinks, examination tables, or food preparation tables becomes an ideal site for infestation because the characteristics of the habitat are similar to that of a tree hole (Fig. 8.4). Leaving some space on the side or behind the equipment will increase airflow, increase brightness, and reduce humidity. Consequently, these pests are denied a dark, damp, tight (protected) harborage. Additionally, this void is now accessible for applying control measures.

Animal feed that is not stored under refrigeration should be placed in containers with tight lids to deny access of larger pests. Bulk material can be divided and placed in smaller storage units so that only one is at risk of contamination at any one time. Adding a piece of dry ice (frozen CO_2) on a paper towel before sealing these will saturate the contained atmosphere with carbon dioxide and

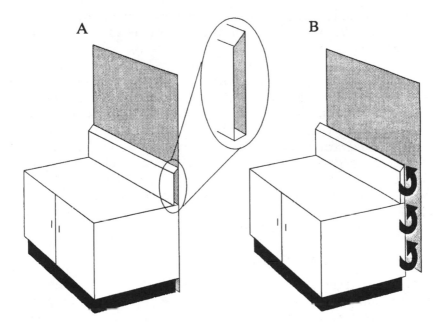

Figure 8.4. Stylized drawing of a common laboratory preparation bench against a wall (A) with end plug removed, revealing a natural void resulting from design of the bench; this area behind the splash board provides a suitable habitat for arthropods, especially cockroaches. Leaving a few centimeters of space between this bench and a wall (B) will increase airflow, reduce humidity, increase brightness, and will allow access for control measures as well.

kill any stored-food pests (mites, beetles, moths) that may be present.

Substrates in Naturalistic Displays

Choice of substrates for a naturalistic display can also affect the likelihood of infestation by unwanted creatures. Mulches play one of the more important roles. Laboratory research has shown that the larger cockroaches preferentially occupy spaces that are 1 cm high (Mizuno and Tsuji, 1974). Logically, mulches with ample interstitial space (e.g., deciduous leaves, pine straw, and pine bark) should function as suitable harborage sites. These substrates provide a matrix within which cockroaches and other invertebrates can move freely in a protected environment. In contrast, mulch made from baldcypress wood pulp compresses, and thereby denies for these relatively large cockroaches a habitat of suitable physical dimensions (Brenner, 1988).

Studies on resting behavior of German cockroaches revealed preferences for significantly narrower harborge spaces ranging from 0.5 to 5 mm, depending on instar and physiological state (fed vs. unfed, gravid; Wille, 1920). Thus, this domestic species, or smaller peridomestic species, conceivably could infest mulches indoors. Recent observations from central Florida on a new pest in the U.S.A., the Asian cockroach (*Blattella asahinai*), indicate that this species readily infests these mulches (R. J. Brenner, unpublished data).[3] Added protection can be gained by mixing tobacco stems and leaves with mulch. This is repellent to some arthropods, and may prevent or retard infestation.

Cockroach Suppression

Specific strategies for controlling cockroaches differ for domestic and peridomestic species, but in either case, the control measures will be directed according to the "CIA" concept. This is where focality and understanding biotic needs of arthropods work to your advantage.

Traps, with the inside upper 3 cm surface lightly greased with petroleum jelly and baited with nontoxic dry distiller's grain (Brenner and Patterson, 1988, 1989), can virtually eradicate a population of larger species of peridomestic cockroaches within a few days. However, traps must be placed in or near principal foci. An alternative is use of toxic bait stations with nonvolatile active ingredients such as hydramethylnon or boric acid preparations. Many of these materials will also be marketed as gels or pastes in dispenser tubes and syringes. This allows application of toxic baits to places such as cracks or voids in doors, table and chair legs, cage frames and supports. Harborage sites that cannot be sealed can be treated with sorptive powders containing various concentrations of amorphous cilica gel and ammonium fluosilicate. These absorb moisture and waxy secretions of the arthropod exoskeleton, further increasing chances of desiccation. Various preparations of boric acid dusts have been commonly used (Rust, 1986), but because boric acid has some toxicity to vertebrates, thin films of this material may be inappropriate in some naturalistic research facilities .

In recent years, juvenoid hormone analogs such as hydroprene (for cockroaches), methoprene (active against fleas), and fenoxycarb (cockroaches, fleas, fire ants) have been developed, which prevent the insect from reaching sexual maturity. Known as insect growth regulators, these materials halt reproduction and the population subsequently is eradicated by attrition (Staal, 1985). Because these materials are specific to arthropods, and often to a narrow taxon, there is little or no risk to nontarget vertebrates. However, because of high reproductive potential in German cockroaches, an initial treatment with toxicants usually is necessary to achieve acceptable levels of control quickly (Patterson and Koehler, 1985; Bennett et al., 1986; Brenner et al., 1988).

CONCLUSIONS

In summary, four points are key to regulating unwanted arthropods in naturalistic research facilities. First, an understanding of the general bionomics of arthropods is essential; water, harborage, and food are the major constraints on survival. Second, in designing structures, anticipate how certain designs or use of certain construction materials may provide these essential factors that predispose such facilities to infestation by arthropods. Third, modify these plans whenever possible to preclude such materials or designs, or, if they are essential to the facilities, install several access points so potential infestations will be accessible. Modify equipment or plan its placement to reduce the suitability of potential harborage sites. Finally, use "CIA" strategies to combat the pests, relying on traps, baits, desiccants, and new-generation endocrine analogs.

ACKNOWLEDGMENTS

I thank D. Milne for assistance in preparing figures, and D. Johnson, Tampa, Florida, R. S. Patterson, Gainesville, Florida, and C. J. Jones, Urbana, Illinois, for helpful reviews.

NOTES

1. Surfer, Golden Graphics Software, Golden, Colorado.

2. The exception to tree species is palm trees; the base of fronds in the canopy provides a suitable habitat. In some palm species, the remnants of old fronds (hollow or spongy inside) remain on the trunk and become infested.

3. The Asian cockroach closely resembles the German cockroach morphologically. However, the Asian cockroach is a strong flier and is attracted to reflected light beginning at sunset. For specific pest potential of this species, see Brenner, Patterson, and Koehler, 1988; or Brenner, 1991.

M. KAY IZARD
MICHAEL E. PEREIRA

9

Design of Indoor Housing for a Breeding and Research Colony of Prosimian Primates

In designing facilities for the maintenance of nonhuman primates in captivity, it is essential to recognize that primates are social animals. Thirty-nine of the 54 primate genera (Richard, 1985), or 70% of extant forms, coalesce into social groups, including two or more adults and their infant and juvenile offspring. Even researchers studying those primates formerly considered to lead "solitary" lives, like orangutans (*Pongo pygmaeus*) and galagos (*Galago, Otolemur,* and *Euoticus spp.*), now stress that these animals are social, and that much remains to be learned about primate social interaction through the study of these species (e.g., Clark, 1978a; Harcourt and Nash, 1986; Rijksen, 1978; Smuts et al., 1986).

The primary adaptations of primates have evolved in environments that include participation in complex social interactions within groups or populations of broadly stable composition. In virtually every species studied to date, between infancy and adulthood, maturing individuals develop strong and typically long-term affinitive relationships with their mothers, siblings, and unrelated conspecifics (Pereira and Altmann, 1985). Affinitive interactions (greetings, grooming, social play, and sleeping associations) balance the agonstic interactions and competition that accompany group living. For primates, normative behavioral development is a life-long process that depends on continuous input from myriad social, as well as ecological, stimuli (Pereira and Altmann, 1985).

Under restrictive living conditions, nonhuman primates cannot be expected to express normal patterns of physiology or behavior. Because animal physiology and behavior are biological systems coadapted to particular physical and social environments invariably present during normal development, radical departures from these environments can be expected to generate aberrant functioning. In fact, a vast literature exists describing effects of housing style and group composition on the physiology and behavior of laboratory animals (Fox, 1986). Over the past 30 to 40 years, research has causally related changes or differences in captive social environments to differential development in animals' brains, musculoskeletal anatomy, neuroendocrine systems, immunological systems, behavior, and reproductive performance (Fagen, 1981; Fox, 1986). Because most primates have sophisticated social systems, valid behavioral research with these species cannot be conducted in sterile, minimalist laboratory environments (Pereira and Altmann, 1983; Pereira et al.,1989). Fox (1986), reviewing generalities and heuristic examples for many laboratory animal species, cogently expands the argument that much behavioral research on subjects maintained in restrictive, highly artificial environments is of dubious validity.

To ensure representative physiology or behavior for a given primate species in captivity, appropriate numbers of conspecifics must reside in social groups, the demographic composition of groups must be species-typical, and group membership must change in a manner characteristic of the species. Space must be provided that is sufficient in size and appropriately structured in three dimensions. For primates to be housed in species-typical social groups, the internal structure of housing must promote species-typical locomotion, vocalization, scent marking, and both affiliative and agonistic social behavior. Fortunately, housing captive primates in representative social groups is not only necessary to conduct valid physiological or behavioral research, but it is also practical (Pereira et al., 1989; Wright et al., 1989).

How then does one go about designing an environment to facilitate both captive breeding of primates and valid research into their behavior and physiology? How can these objectives be met while maintaining high standards of animal health and retaining realistic management procedures? To provide some answers to these questions, we will describe in detail the indoor facilities in which nocturnal prosimian primates and two large-bodied, diurnal species of lemurs are managed at the Duke University Primate Center (DUPC). To succeed, design considerations clearly must accommodate the needs of the animals, as well as accommodating management considerations and the objectives of researchers. The approach presented is economical and would be applicable in almost any existing laboratory environment.

THE DUKE UNIVERSITY PRIMATE CENTER

The Duke University Primate Center is a research and captive breeding facility dedicated to the conservation of prosimian primates and to the training of conservationists, researchers, and animal husbandry personnel for work with these species (Izard, 1989). Prosimians comprise seven of the 14 families within the Order Primates, and include the Lemuridae and Lepilemuridae (true lemurs), Cheirogaleidae (mouse and dwarf lemurs), Indriidae (indris, avahi, and sifakas), Daubentoniidae (aye-aye), Lorisidae (lorises and galagos), and Tarsiidae (tarsiers). All extant lemuriform species (the first five families above) are found only in Madagascar and are listed as endangered species by the International Union for the Conservation of Nature (IUCN, 1988). Many of these species are rarely sighted in their homeland (Jolly, 1986; Pollock, 1986a). The DUPC houses the only captive populations of several prosimian species (Pollock, 1986b; Tattersall, 1982).

Habitat destruction and the threat of extinction for several prosimian species over the next 10 to 15 years has led to international recognition of the DUPC collection as a critical second line of defense for protection of these species (Jolly, 1986; Pollock, 1986a,b; Tattersall, 1982). The DUPC has an unrivaled record of propagation of several of the most difficult prosimian species to rear in captivity (Table 9.1), and now maintains more than 500

TABLE 9.1
Infant Births and Survivorship at the DUPC, 1960–1989

Family Genus Species	First year at colony[1]	Total number of births to date	Survived first week (percent)	One week- olds that survived first 6 months (percent)	Overall Survival to 6 months (percent)
Lemuridae					
Lemur					
fulvus	1960	550	79	92	73
catta	1962	150	91	96	87
macaco	1968	111	77	93	71
Mongoz	1976	35	83	90	74
coronatus	1978	29	72	86	62
rubriventer	1979	5	100	100	100
Varecia					
variegata	1962	265	82	89	73
Hapalemur					
griseus	1970	20	75	100	75
Indriidae					
Propithecus					
verreauxi	1968	26	65	71	54
Cheirogalcidae					
Cheirogaleus					
medius	1969	215	71	95	68
major	1984	0	—	—	—
Microcebus					
murinus	1969	183	87	90	79
Mirza					
coquereli	1982	79	68	96	66
Tarsiidae					
Tarsius	1985	20	15	100	15
Lorisidae					
Galago					
senegalensis	1960	241	75	78	59
demidovii	1967	9	56	100	56
Otolemur					
crassicaudatus	1961	320	68	84	57
garnettii	1959	250	73	93	68

[1]Colony began at Yale University in 1960 and moved to Duke University in 1966.

TABLE 9.2
Recent DUPC Research on Behavior, Physiology, and
Reproductive Biology of Prosimians

Topic	Reference
Behavior	Hoffman & Foerg, 1983
	Glander & Rabin, 1983
	Macedonia & Taylor, 1985
	Taylor & Sussman, 1985
	Ganzhorn, 1986
	Wright et al., 1986b
	Macedonia, 1986, 1987
	Gebo, 1987
	Pereira et al., 1987
	Cherry et al., 1987
	Kappeler, 1987a
	Pereira et al., 1988
	Vick & Pereira, 1989
	Macedonia & Polak, 1989
	Dubois & Izard, 1990
	Kappeler, 1990
	Pereira et al., 1990
	Milliken et al., 1991
	Pereira & Weiss, 1991
	White, 1991
Physiology/Reproductive Biology	Foerg, 1982
	Izard & Rasmussen, 1985
	Rasmussen, 1985
	Izard & Simons, 1986
	Wright et al., 1986a
	McNab & Wright, 1987
	Kappeler, 1987b
	Rasmussen & Izard, 1988
	Izard & Fail, 1988
	Pereira & Izard, 1989
	Hughes et al., 1990
	Izard et al., 1991
	Pereira, 1991
	Perry et al., 1992

individuals representing 29 prosimian taxa. In addition, the DUPC is the only institution where research into the behavior, physiology, and reproductive biology of diverse prosimians can be conducted (see Table 9.2).

Resident nonhuman primates have traditionally been housed in social groups under naturalistic conditions, to meet the DUPC goals of enhancing captive propagation while fostering research. Over half of the colony members are housed outdoors. Surrounding the main building are more than 80 runs or cages in which social groups of diurnal lemurs live year round. These cages enclose between 25 and 250 cubic meters of space, and each has a wire-covered top, in lieu of a roof, to allow the lemurs to experience natural patterns of sunlight and weather. Each outdoor run includes one or more elevated, heated boxes to provide refuge from severe weather conditions. In addition, each run contains naturalistic substrates such as trees, vines, and bamboo. The DUPC also maintains 12 social groups of lemurs, representing eight species in multihectare forest enclosures (Pereira et al., 1987, 1988, 1989, 1990; Pereira and Izard, 1989; Taylor and Sussman, 1985).

While most of the diurnal, large-bodied lemurs housed at the DUPC are maintained outdoors (or in indoor-outdoor housing), the small-bodied nocturnal mouse and dwarf lemurs (Cheirogaleidae), tarsiers (Tarsiidae), and lorisiform colony members (Galagidae and Lorisidae) are housed socially in indoor rooms outfitted with naturalistic substrates (Wright et al., 1989). These species are much smaller than the diurnal lemurs, and therefore can be housed in smaller quarters. With the exception of the aye-aye (*Daubentonia madagascariensis*), the nocturnal prosimians weigh less than 2.5 kg; the smallest, the mouse lemur, only weighs 60 g. Reasons for not housing these species outdoors include their small size and vulnerability to predation, as well as the difficulties of caring for and studying captive nocturnal animals without at least partially shifting their light cycle. Even though the nocturnal species are less gregarious than the larger diurnal taxa, they still present the challenge of providing appropriate housing that meets their limited social needs. They also require housing that provides them with hiding places and the opportunity to escape from or get out of sight of their cagemates.

Design Considerations for Animals

Tremendous diversity exists among prosimians with regard to size, locomotion, diet, reproduction, and social system. The indoor housing environments at the DUPC address specific needs generated by this phyletic diversity. Many rooms are designed exclusively for a particular species. All provide for the basic necessities of feed-

Figure 9.1.   Aye-aye (*Daubentonia madagascariensis*).

ing, exercise, and resting, and provide a complex environment that is designed to enhance the quality of life experienced by the species in captivity. As many as 13 nocturnal prosimian species have been housed in indoor rooms suitable for conducting behavioral research (see Table 9.3). In addition, because of their rarity, two diurnal species, the grey gentle lemur (*Hapalemur griseus*) and the sifaka (*Propthecus verreauxi*), have been housed in indoor rooms with access to outdoor caging.

Naturally occurring plant species are used for furnishing the indoor rooms. These include bamboo, cedar trees, grape vines, kudzu vines (*Pueraria lobata*), and various branching trees. All are preserved with several coats of nontoxic polyurethane, which facilitates cleaning while maintaining the natural texture best for locomotion, sleeping, and scent marking. For the animal's safety, these substrates are anchored firmly with heavy wire to the walls, floors, or ceilings of rooms. Small trees or bushes are set in concrete. Larger trees are secured both to the floors and walls. Care is taken to eliminate sharp, protruding ends of wire that might harm an animal. Nest boxes or hollow logs used for resting and sleeping are provided for respective taxa. Nest boxes are elevated for the arboreal prosimians, as they prefer to sleep as high off the ground as possible. Elevated shelves for resting are provided for animals that do not use nest boxes (i.e., *Propithecus*, *Tarsius*, and *Hapalemur*).

TABLE 9.3
Prosimian Primates Housed in Indoor Naturalistic
Facilities at the DUPC

Scientific Name	Common Name
Tarsius bancanus	Bornean tarsier
Tarsius syrichta	Phillipine tarsier
Loris tardigradus	Slender loris
Nycticebus pygmaeus	Pygmy slow loris
Nycticebus coucang	Slow loris
Perodicticus potto	Potto
Otolemur crassicaudatus	Large-eared greater galago
Otolemur garnettii	garnett's greater galago
Galago moholi	Mohol lesser galago
Microcebus murinus	Lesser mouse lemur
Mirza coquereli	Coquerel's mouse lemur
Cheirogaleus medius	Fat-tailed dwarf lemur
Cheirogaleus major	Greater dwarf lemur
Daubentonia madagascariensis	Aye-aye

The indoor environments range in size from 3 m^3 to 82 m^3. The smaller rooms have been used to house the smallest species, such as the mouse lemur (60 g), tarsier (115 g), and slender loris

(200 g). The larger rooms house the aye-aye and the two large-bodied diurnal prosimians—the grey gentle lemur (1–2 kg) and the sifaka (4–6 kg).

In orienting substrates, a species' use of space in three dimensions is the primary consideration. Mouse lemurs are small arboreal quadrupeds with limited jumping abilities. Their rooms are furnished with small tangled vines (honeysuckle or grape vines work well) that provide continuous runways between feeding and resting areas. Rooms for the larger, fat-tailed dwarf lemur (300 g), another arboreal quadruped, are furnished with vines of a larger diameter in which the pathways are more open to accommodate the larger animal. For species weighing more than 1 kg, still larger vines and branches are used. For the tarsier, a small-bodied vertical clinger and leaper, rooms are furnished with many vertical-oriented branches. Bamboo is particularly well suited for use as a vertical substrate and is commonly used. For the 3–4 kg sifaka, another vertical clinger and leaper, sturdier vertical substrates (usually cedar trees) are spaced several meters apart. For all species, optimal arrangement of locomotor substrates not only increases available space by using a room's internal space, but also provides multiple routes for escape from aggression. Locomotor pathways, vine tangles, and hollow logs provide opportunities for visual isolation from cagemates.

Important safety considerations include prevention of animal access to light fixtures and fans. Light fixtures are shielded with vinyl-coated wire and, if necessary, covered overhead with plastic to prevent urine and feces from falling on them. Florescent light fixtures are usually recessed but, in case they are not, a barrier is installed to prevent an animal from crawling above the fixture. Fans for ventilation are recessed and covered with small wire mesh.

Finally, ease of access to food and water is critical. For example, while most prosimians will drink from water bottles with sipper tubes, some (i.e., *Propithecus*, *Tarsius*, and *Loris*) will not. Fresh water is supplied daily to these species in a shallow container. Arboreal pathways to food and water are provided in a manner sensible for the particular species.

Design Considerations for Management

Housing characteristics at the DUPC accommodate not only the needs of particular species but also the needs of management per-

Figure 9.2.   Golden-crowned sifaka (*Propithecus tattersalli*). Note the vertical cedar tree in background with mango leaves and redbud cuttings attached. Giant bamboo and grape vines are oriented at varying angles.

sonnel. Indoor animal rooms are designed to facilitate cage cleaning, feeding, visual inspection, and animal capture and handling. In the indoor rooms, either wood chips or sand is used to cover floors which are either washable vinyl or concrete. The floor coverings not only minimize odors, but also provide a cushion when an animal falls. Wood chips are absorbent, inexpensive, and easily stored when purchased in bags. However, wood chips are inappropriate for tarsiers, which eat only live animal prey (Neimitz, 1984), but will not search through chips for hidden prey items (crickets and lizards). Therefore, sand is used in tarsier rooms.

Feces and wet spots in sand or wood chips are removed daily. In many cases, the inhabitants of an indoor room defecate consistently at the same location, which simplifies daily cleaning. At regular intervals, animals are detained in holding cages, old bedding is discarded, and rooms, including furnishings, are thoroughly washed with disinfectant, sprayed for cockroaches, washed again, and refurbished and repaired, if necessary. Several times a year, cage furnishings are replaced with fresh material. Ideally, new substrates are arranged, using the same techniques, in different arrays to provide environmental novelty within rooms.

Prosimian primates depend heavily on olfactory means of communication (Schilling, 1979). Provision of natural materials for scent-marking encourages normal scent-marking behavior and promotes research on olfactory communication. Although wood cannot be sanitized to the extent of stainless steel, it is much more appropriate for rooms in which behavioral observations will be conducted. More than 20 years of experience with wood-based substrates in animal rooms at the DUPC has provided no evidence that, maintained as described above, it harbors pathogens that jeopardize the welfare of prosimian primates.

Animals are visually inspected at least once (and usually twice) daily to detect any medical or behavioral problems. With nocturnal primates, this is easiest during the inactive period while the lights are on. Nest boxes with open fronts or Plexiglas sides allow viewing the animals without handling. However, some problems may be missed if the animals are not observed when they are active. Timing the light cycle so that active and inactive periods both occur during the normal working day ensures that the animals can be cared for optimally. Moreover, cleaning can be done during the inactive period, while observations can be continued during the active period.

The environment is also designed to facilitate the capturing of the animals by the technicians. Nocturnal prosimians that use nest boxes are caught during their inactive periods by trapping them within the nest box. A mechanism for closing off the front of the nest box is useful for wary, fast-moving animals and minimizes chasing of the animals. Interior furnishings are located to allow technicians ease of movement within rooms in case animals need to be netted.

It is imperative to provide nest boxes for animals that normally give birth in nests, as do many of the nocturnal prosimians. In the absence of a suitable nest box, neonatal mortality is unnecessarily high (Katz, 1980). A one- or two-inch lip across the front of the nest box prevents infants from falling out.

Food dishes are anchored to prevent tipping, and are placed near room entrances with shelves underneath to catch dropped food items. Most nocturnal species initially pick up and drop food items to the floor or shelf below in searching their food dish for preferred items. These dropped items will be retrieved by the animal later in the active period, after the preferred food items have

Figure 9.3. Gentle lemur (*Hapalemur griseus*). Bamboo forage is attached to a cedar tree. Side branches have been left on the tree for perching.

been consumed. This arrangement not only conserves food, but also provides the animals with foraging opportunities. Folivorous prosimians are provided fresh browse (leafy branches, bamboo, mango leaves) in containers of water to keep the browse hydrated. One or more containers are placed in semipermanent holders bound to tree trunks, branches, or vines. Although the content of some diets is varied daily (or at slightly longer intervals), no attempt is made to vary either the location of the food or the manner in which it is presented. The animals appear to become accustomed to receiving their diet in a consistent manner, but become confused if the routine is varied. Depending on the species, daily diets might include Purina High Protein Monkey Chow (Ralston Purina), a wide range of fruits and vegetables, foliage, crickets, mealworms, lizards, or grubs.

Design Considerations for Researchers

Research at the DUPC is primarily behavioral, but the physiology and reproductive biology of prosimian taxa are also studied. To collect meaningful data of any kind, naturalistic behavior, including social interactions, must be made possible. Social groups of species-typical composition, housed within naturalistic physical environments, allow for expression of normal patterns of behavior.

A primary consideration, of course, is the ease with which behavior can be observed. From the observer's standpoint, most parts of a room should be well within view. At the DUPC, some indoor rooms are arranged in pie-shaped wedges around a central observation core, so that three to five rooms can be observed from a single location. Lighting conditions must allow good visibility but not disrupt the animals' behavior. For study of nocturnal prosimians, red lamps come on automatically when the white lights go off. Another consideration is the time of observations. As mentioned earlier, light cycles are partially reversed for nocturnal animals, such that the dark phase (scotophase) occurs between midday and midnight, and the light phase (photophase) between midnight and midday. Thus, feeding and cleaning can be done early under white light and behavioral sampling can be conducted later in the afternoon and evening.

CONCLUSIONS

We have addressed general issues and described specific solutions to problems of establishing and maintaining naturalistic indoor environments for prosimian primates. At the DUPC, the needs of the animals, technicians, and researchers are accommodated. This has resulted in successful programs of captive breeding (Table 9.1), as well as a productive research program (see Izard, 1989; Table 9.2).

Many existing standards for indoor environments supporting animal research actually preclude naturalistic behavior, thus creating conflict between the goals of the researchers and requirements of the Animal Welfare Act, as enforced by the United States Department of Agriculture (USDA). For example, if preoccupation with the sanitizability of indoor rooms were to culminate in a mandate for exclusive use of stainless steel or plastic tubing for locomotor pathways, instead of natural materials, then, based on our observations of substrate use, increased rates of behavioral pathology, infant death, and adult injury from falling could be expected in the DUPC colony of nocturnal prosimians.

In an ironic synergy, current pressure from the animal welfare movement, in tandem with traditional concepts of health maintenance in colonies of "laboratory animals," has unwittingly fostered an intitial reaction against naturalistic environments and toward sanitizable cages with furnishings of plastic and steel. Whereas animal welfare advocates have sought improvement in animal housing conditions, this initial response by regulatory agencies clearly and unfortunately encourages the opposite development, especially for nonhuman primates .

In its new mandate, the USDA requires laboratories to provide for the psychological well-being of their animals. From more than 20 years of behavioral observation of DUPC's diverse collection of captive prosimian primates in the world, it is known that the nature of psychological well-being may differ substantially among species. Psychological well-being can be meaningfully estimated only through a careful, stepwise process involving a rigorous methodology. First, it involves extensive behavioral observations in naturalistic settings and study of the available literature. The data should suggest which modifications to housing and social environments will promote psychological well-being. Collection of empirical data from well-designed experiments should follow. Adjustments to one group should be followed by further observation and comparisons to a control group.

This process will benefit tremendously from the techniques, advice, and ideally, supervision of experts in animal behavior and ecology. Our hope for the future is that regulatory agencies will work closely with scientists whose careers are focused upon the documentation and examination of the naturally occurring behavior and ecologies of animals. In this scenario, scientists who maintain animals in captivity will be empowered to do everything possible to provide conditions that simultaneously accommodate their subjects' evolved needs and validate their own research.

ACKNOWLEDGEMENTS

We would like to thank the former director of the DUPC, Dr. E.L. Simons, for providing the opportunity to work with the prosimian colony. Dr. J.G. Vandenbergh made helpful suggestions to improve an earlier draft of this manuscript. We gratefully acknowledge the assistance of Louise Martin, who compiled the data in Table 9.1, and Steve Daugherty who took the photographs. Paper no. 459 of the DUPC.

JAMES G. DOHERTY
EDWARD F. GIBBONS, JR.

10

Managing Naturalistic Animal Environments in Captivity

The management of naturalistic animal environments in captivity involves everything from architectural design to daily maintenance and security, to keeper work efficiency. Good management begins when species are chosen for the facility, and theoretical and pragmatic aspects of design are discussed. Management continues through construction to include the introduction of animals, the long-term maintenance of the facility, the animal collection, and the reuse or destruction of the facility. Thus, the management of naturalistic animal environments can be divided into six phases: selection, information, design, implementation, use, and reuse or demolition (e.g., Fisk and Lewis, 1982). The phases are interrelated, and good management requires the continual coordination and modification of administrative strategies within and between each phase. The purpose of this chapter is to detail a few of the management considerations involved in each of the six phases of facility design and operation. Discussion will begin with a few rationales for developing naturalistic environments in captivity.

RATIONALE FOR NATURALISTIC ENVIRONMENTS

In order to effectively manage naturalistic animal environments there must exist a general administrative philosophy, and central to the development of such a philosophy is the rationale for the existence of these facilities. In general, there are four reasons for building naturalistic environments: (1) to provide for the ethical treatment of animals, (2) the need to maintain and breed wildlife in captivity, (3) to enhance the scientific study of animals, and (4) public education. Any number and combination of these rationales can be applied to a given facility.

Ethical Treatment of Animals

Scientific advancement in fields such as animal behavior, ecology, evolution, physiology, and veterinary medicine have provided important insights into the biological and psychological requirements of animals (Curtis, 1985; Hafez, 1968). Research has indicated that failure to provide for these needs can lead to deterioration in the animals' physiological and behavioral functioning (Ader, 1967; Besch, 1985; Goosen, 1988; Stoskopf, 1983; Weihe, 1971). Deficiencies in the physiological and behavioral functioning of animals in captivity may affect the internal and external validity of scientific findings derived from these subjects (Weihe, 1988). Consequently, from ethical as well as scientific perspectives, we are bound to develop captive environments that meet the requirements necessary for the biological and psychological well-being of animals.

The Need to Maintain and Breed Wildlife Captivity

Many of today's wild populations are becoming endangered as a result of habitat destruction, poaching, and illegal trade (McNeely et al., 1990; Western and Pearl, 1989; Wilson, 1989). As wild populations continue to decline there will be an increased need to maintain and breed animals in captivity for the purpose of reintroducing them into the wild. The maintenance, breeding, and reintroduction into the wild of exotic animals, however, is not an easy process and requires the implementation of health care and husbandry programs to ensure that animals are free from disease and are physiologically suitable for such programs (Brambell, 1977). In addition, it is imperative that the animals learn species-typical behavior patterns that will enhance the probability of breeding success as well as survival in the wild (Kleiman, 1989).

Traditional research laboratories and outdated zoo environments generally do not provide for the variety of environmental and social experiences necessary for the most effective husbandry, breeding, and reintroduction programs. These environments do not "encourage" animals to express the wealth of behaviors typical for the species in the wild, and in some cases promote the occurrence of abnormal behaviors that may be detrimental to the animals' health and well-being (Fox, 1968; Goosen, 1988; Hediger, 1969).

In contrast, because of the general complexity of naturalistic environments, animals are afforded the opportunity to display a wider range of behaviors in manners appropriate for the species. In these environments animals can better interact with conspecifics from a variety of age/sex classes and, in the case of mixed-species enclosures, with individuals from other species. The variety of behavioral and social experiences available to animals in naturalistic environments will increase the likelihood of breeding success, and will make them good candidates for reintroduction into the wild.

Scientific Study of Animal Behavior

A third reason for building naturalistic environments is that they can provide the opportunity for scientists to investigate aspects of animal behavior that may be difficult or impossible to study in the wild. Because naturalistic facilities can approximate wild habitats with respect to selected environmental variables, researchers are afforded unique opportunities to experimentally test hypotheses regarding the ecological, evolutionary, and social importance of behaviors (Hediger, 1969; Janson, chap. 18, this volume; Menzel, 1969). In these environments, scientists may elucidate the physiological, morphological, and cognitive mechanisms that drive behavior.

Education

A fourth justification for building naturalistic environments is that they can serve as instruments to educate people about animals and ecosystems, the need to conserve wildlife and their habitats, and to foster a general appreciation for the wonders of the animal and plant kingdoms (Gibbons, 1994). An education program should be a vital component of every naturalistic environment.

CONSIDERATIONS IN THE MANAGEMENT OF NATURALISTIC ENVIRONMENTS IN CAPTIVITY: SIX PHASES OF FACILITY DESIGN

The rationale for building naturalistic environments in captivity will serve as the philosophical cornerstone that will guide the development of management objectives in each of the six phases of facility design and operation. It is from this interface between rationale and objectives that management decisions will follow a course of continuity and logic within and between each of the six design phases.

Selection Phase

The selection phase of facility design involves decisions regarding the selection of professionals who will comprise the facility and animal management team, and what species to house in the environment. Outlined below are a few of the criteria that should be considered during each of these selection processes.

Facility and Animal Management Team (FAMT). The selection of a facility and animal management team is critical for the successful design, construction, and operation of the naturalistic facility and the health of the animals. Because of the complex interactions between environmental design, management, and animal health, it is imperative that the focus of the team be multidisciplinary. The disciplines represented on the team should be periodically reviewed and modified according to need. Among the professionals who should be asked to participate on the FAMT are administrators, animal behaviorists, animal technicians, architectural engineers, behavioral ecologists, curators, and veterinarians. The valued contributions from each of these specialists will help to ensure the success of the overall design process, the effective management and use of naturalistic environments, and the long-term health and well-being of the resident animals.

Species Selection. In a research facility probably the most important criterion in species selection is the scientific interest and expertise of the principal investigators. Beyond this, consideration must be given to the conservation status (endangered, threatened) and availability of species; the number and compatibility of animals and species to be included in the environment; legal aspects involving the importation of species from either the wild or another captive facility (especially with regard to endangered species); institutional support for the animals and facilities; and ease of deposition of the animals upon termination of the research program. It is imperative that sound management and scientific justifications be given for the selection of species, and that the justifications be properly documented with the institution's animal care and use committee.

Information Phase

It is in the information phase of facility design that a multidisciplinary FAMT will be worth its weight in gold. Specifically, the objective is for the team to gather information related to questions concerning the design, construction, use of the facility, the biological and psychological requirements of the respective species, and the scientific goals of the facility (Fisk and Lewis, 1982). The expertise of each team specialist will help to ensure the collection and synthesis of information that will address questions related to each of the above mentioned and related topics. Table 10.1 lists a few of the questions that should be addressed during the information phase.

TABLE 10.1
Considerations to Be Researched During the Information Phase
of Facility Design

1. What are the exhibit and holding area spatial requirements for each designated species?
2. What type habitats would these species occur in nature, and how well can we attempt to duplicate them?
3. Can we meet the dietary needs of the animals?
4. Can we mix species?
5. What barriers are needed to confine these species safely and unobtrusively?
6. How can we best separate animals from observers?
7. How much daily maintenance time will the animals and their enclosures require?

Design Phase

The success of the design phase is largely determined by the planning and outcome of the two previous phases. If careful consideration has been given to the composition of the FAMT, the selection of species for the facility, and the gathering and synthesis of information, there will be fewer problems during this critical stage in the development of the environment. During the design phase, the FAMT should also keep in mind the implementation, use, and reuse or demolition phases of design. Finally, the initial design of any naturalistic environment in a research facility should be flexible enough to permit the future remodeling of the facility, thus avoiding demolition (Graves, 1990). Naturalistic facilities in zoological parks provide excellent examples of the use of flexibility in design. The following will highlight a few of the many issues that can become factors during the design phase.

Human versus Animal Needs. The issue of human versus animal needs in design is a management problem basic to all naturalistic environments in captivity. This conflict frequently exists because the actual needs of humans and animals may not be totally clear, and the respective needs are viewed as opposing and mutually exclusive. The effective resolution of this conflict is vital to the operational success of naturalistic facilities.

One method of resolving conflicts between human and animal requirements involves the independent listing of the respective needs (Wallace, 1982). A comparison of these lists will define what needs are in conflict (as well as those that are compatible). Once this is evident the FAMT can begin to seek solutions through compromises in the design plan (Wallace, 1982). However, regardless of the method used to resolve conflicts in design, a basic principle should always hold true: animals' needs have priority. This principle should also be applied to the actual operation of the facility, including research.

Replicating Design Characteristics of Other Facilities. In an era of escalating design costs and diminishing budgets, it is all too tempting to replicate designs that are already in use. While it is advantageous to copy facility design features that have passed the test of "real-time" operation, it is equally important to avoid the reproduction of faulty design features (see Hediger, 1969, pages

183–216, for an excellent overview of this topic). The uncritical replication of inappropriate facilities will inhibit the creative growth of environmental design for animals, and may affect the animals' health and well-being (see Stoskopf and Gibbons, chap. 11, this volume).

Geometric Shapes. All too frequently, captive environments are designed so that the horizontal and vertical space consists of a series of straight lines, right angles, and assorted cubical or rectangular forms. The inclusion of such geometric shapes in design usually translates into reduced construction costs. However, as Hediger (1969, p. 21) pointed out: "In . . . architecture it is not the simplest and the cheapest type of building which should be given primary consideration but the type of building which comes nearest to meeting the [animals'] biological requirements. The cube, indeed a straight line of any kind, is unbiological."

In designing naturalistic captive environments every attempt should be made to contour substrates and surfaces to approximate environmental characteristics that the animals would encounter in nature. It is important that such an approximation take into account any special physiological, morphological, behavioral, and cognitive features of the animals (e.g., Jouffroy and Lessertisseur, 1979).

Computers and Research Space. Recent technological advances in computer and software design can be used to develop model environments that will enable the FAMT to "view" the facility and to detect and correct design flaws prior to the formalization of plans. The application of computers to the design of naturalistic facilities may also enable the FAMT to simulate "real-time" operation of the facility, including the behavior of the animals. The creative use of computerized models can lead to advances in environmental design not thought possible with traditional drawing and scale model representations.

The importance of computers in the design phase does not end with the development of working models. The actual design should include a computer room from which environmental parameters such as lighting, temperature, and humidity may be regulated and monitored (Bancroft, 1985; Lofgreen, 1987). The use of computers to regulate environmental parameters may permit the development of microhabitats that will increase the "naturalness" of the captive environment. The use of computers will also permit the easy collection and analysis of valued environmental information. The availability of such data will be an asset to the management of naturalistic facilities, and to the health care of the animals (see Stoskopf and Gibbons, chap. 11, this volume).

The design phase should also include research space that for convenience and budgetary reasons might be included in the above mentioned computer room. This aspect of design, however, extends beyond the inclusion of research rooms where files can be stored and data analyzed. Research design should incorporate observational blinds in the naturalistic environment, and the strategic placement of video monitors so that technicians may monitor the animals without having to enter the environment.

Animal Containment. Animal containment is one of the most basic concerns in the design of naturalistic facilities. Problems were few when bars, wire, and high fences were used to contain animals. Now glass, moats, man-made cliffs, and mud banks are also used to keep animals in their enclosures. While these containment devices enhance the visual aesthetics of the environment, they create a multitude of problems not present with more traditional barriers (Hancocks, 1971; Hediger, 1969). Specifically, the use of naturalistic barriers can create escape routes that are a result of the animals' ability to more fully express their behavioral capacities in naturalistic environments. Thus, animals may engage in feats of behavior that were thought improbable, if not impossible. Consequently, when designing naturalistic barriers it is important to take into account not only what animals normally do, but also what they are capable of doing when properly motivated. The potential capability of animals' behavior can often be assessed through interdisciplinary studies involving the physiological, morphological, behavioral, and cognitive characteristics of the species or similar taxa (e.g., Prost, 1965; Jouffroy and Lessertisseur, 1979).

The use of naturalistic barriers within an environment can provide the illusion that predator and prey are together in the same enclosure. Naturalistic barriers can also provide an area of refuge when animals experience interspecific or intraspecific conflict, may keep animals from accessing certain parts of the enclosure (e.g., tree trunks), and may enhance daily management and veterinary care procedures. The specific uses of naturalistic and other barriers will depend on the multitude of ways they can enhance the health and well-being of animals in captive environments.

Animal Behavior. It is in part because of an interest in having animals engage in as much of their normal range of behaviors as possible that naturalistic environments are designed. The management of behavior, however, is one of the most important and perplexing problems facing the FAMT. Of importance is the resolution of conflict that may arise from the need to manage animal behavior for husbandry and veterinary care purposes, and the impact such methods may have on behavioral research. As stated earlier, the guiding principle to the resolution of this conflict is that every effort should be made to ensure the animals' health and well-being. The promotion of effective dialogue within the FAMT will, in most cases, lead to solutions that will serve the needs of animal care and research.

There are, however, simple design features that can be used to help manage the behavior of animals, and to promote valid animal behavior research. One of these features involves the inclusion of foraging stations in the environment that will enable animals to "work" for their food (e.g., Carlstead et al., 1991; Lindburg, 1988; Markowitz and Woodworth, 1977). The foraging stations can be as simple in design as food troughs that can be easily cleaned, or as complex as timed or computer-operated devices that distribute food at preselected intervals. In either case the strategic design of foraging stations in the environment can allow for more typical foraging behavior, decrease boredom and some aberrant forms of behavior.

Multispecies Environments. Another effective way to manage animal behavior is to house more than one species in the environment. It goes without saying that the species in a multispecies environment must be physiologically and behaviorally compatible. Two case reports will highlight ways in which multispecies environments can enhance animal behavior. The first involves a simulated mangrove swamp in JungleWorld at the Bronx Zoo, where a troop of proboscis monkeys (*Nasalis larvatus*) is housed with a family of small-clawed otters (*Amblyony cinerea).* Investigations indicated that inclusion of the otters in the exhibit provided a source of some social stimulation for the proboscis monkeys. The introduction of the otters also increased the monkeys' usage of the environment in a manner that was ecologically appropriate for the species. In particular, there was an increase in the arboreal behavior of the juvenile monkeys (Koontz, personal communication, 1991). It should be noted that these two species "use" different parts of their environment, and that they often rest at different times during the day. These facts have made the mangrove swamp the most popular and successful exhibit in a very popular building.

The second case, reported by Stoskopf (1994), involves predator-prey interactions in fish. In this example, a number of compatible fish species were introduced into a large aquarium with artificial reefs. The fish failed to settle into the tank, and swam continuously and refused to eat. After several weeks fish losses due to exhaustion and disease were severe, and necessitated the inclusion of a larger predatory fish species in the environment. The effect of this manipulation was quite dramatic, and as stated by Stoskopf (1994): "The impact was nearly instantaneous. Exhausted prey fish immediately took cover and began to settle into their system, avoiding the predator's territory. They also began to eat and sleep normally. Multistationed feeding has made this arrangement a long-term success for more than 7 years."

In general, there are a few behavioral questions that should be considered in the design of mixed-species enclosures:

a. Are there species-specific requirements that need to be met? Will these requirements present problems from species and management perspectives?
b. How will the species be introduced to the exhibit and to each other?
c. Is the off-exhibit holding area adequate for the respective species?
d. Can the species be fed separately?
e. Will breeding or nesting on the part of any of the species result in interspecific or intraspecific social problems?
f. Is there any possibility that hybridization will occur between the species?

Implementation Phase

The implementation phase of facility design involves the development of a budget and the actual construction of the facility (Fisk and Lewis, 1982). In developing a budget, it is essential that funding accurately reflect construction costs and expenditures associated with "real-time" operation of the facility. Of further importance is the inclusion of budgetary items central to the well-being of the animals (e.g., food of adequate quality and quantity, and proper medical care). In general, the more specialized the physiology, morphology, and behavior of a species the greater the cost to build and properly maintain naturalistic environments, and to ensure the inhabitants' well-being. This principle, holds true for invertebrates as well as vertebrates (Drummond,1994; Perry-Richardson and Ivanyi, 1994).

The budget for building and maintaining naturalistic environments should also include adequate funding for support facilities. These areas include night holding areas, shift cages, diet preparation rooms, general work areas for the staff, and research and computer space. These behind the scenes areas are the "lifeline" to naturalistic facilities. It is from these facilities that some management decisions are made and carried out that affect the naturalistic environment, including the animals. It is unfortunate, however, that when construction and maintenance expenditures exceed original cost estimates, these support areas are often the first to be reduced or cut from the overall facility. Cost estimates in building and maintaining naturalistic environments should ensure that there be adequate funds for support facilities, and funding set aside for the future modification and upgrading of these most necessary areas.

Another cost to managing naturalistic facilities involves the hiring and retention of qualified personnel to care for these environments and the animal collection. The effective day-to-day management of naturalistic environments requires a highly educated and motivated staff. It is difficult to hire and retain such personnel without an adequate salary structure, including fringe benefits. It is also necessary for personnel to keep abreast of technological and scientific advances in the multitude of disciplines associated with the management of animals in captivity. Consequently, funds should be set aside so that staff can attend scientific conferences and seminars related to the species they manage. Failure to allocate sufficient funds for these matters may have a negative impact on staff morale and their ability to provide optimal care for the animals. Insufficient funding may also make it difficult to retain qualified personnel. Low staff motivation along with a high employee turnover rate may have profound negative effects on the management of naturalistic facilities, and the health and well-being of the animals.

Construction. The construction of the facility should include the active participation of the FAMT. It is imperative that the team be on site to ensure that every aspect included in the design plans is adequately built into the environment. Invariably there will be problems during the actual construction of the facility, and the FAMT must be available to effectively resolve these issues. Any delay in addressing construction problems will result in an increase in budget costs, and a potential delay in the opening of the facility.

During the construction process it is important to avoid having cracks and crevices that do not drain. In general, do not accept any rock work where water can lie in puddles. The same holds true for trees and vines. Any work that will allow water to puddle up in crotches or cracks may affect animal health. It is also imperative that any structures built into the environment be able to support the animals. Pools and water moats should have a number of places for animals to easily get into and out of the water, and the water need not be any deeper than absolutely necessary for animal containment or typical behavior.

Use Phase

The use phase begins with the animals' occupancy of the environment, and involves the evaluation of the facility during "real-time" operation (Fisk and Lewis, 1982). Consequently, if there is one word that best describes the use phase of facility design, it is *performance: performance* of design features, *performance* of the animals, and *performance* of the staff. The effective interaction between all three performance factors will ensure the postoccupancy success of the naturalistic environment.

Performance of Design Features. Concern with design features does not end with the construction of the facility. Once the animals are introduced into the environment, the design features must be regularly monitored and evaluated to ensure animal as well as human safety (Stoskopf and Gibbons, chap. 11, this volume). The postoccupancy monitoring of environmental quality will help to ensure that all design features are operating at peak efficiency which will help reduce operational costs (White, 1982).

Performance of the Animals. The monitoring of animal behavior by the FAMT is an important aspect in the postoccupancy evaluation of a facility. Behavioral performance of the animals can point to design flaws in the facility, and may serve as an early indicator of declining animal health and well-being (Gibbons and

Stoskopf, 1989). The rigorous and systematic collection and evaluation of behavioral data should be an important tool in the management of animals in naturalistic facilities (Eisenberg and Kleiman, 1977; Hutchins, 1988).

Performance of the Staff. The postoccupancy success of any naturalistic animal facility relies heavily on the ability of the animal staff to maintain the facility and care for the animals. Consequently, an important component in the management of naturalistic environments involves the effective supervision, evaluation, and training of staff (Cummings, 1986; Faulkner, 1989; Hayden, 1987). It is imperative that any deficiencies in staff performance be promptly reported, and corrected in a constructive manner. The appraisal process should encourage increased productivity and quality of work (Hayden, 1987).

Reuse or Demolition Phase

As happens with any man-made structure, naturalistic animal environments will eventually either have to be renovated or demolished depending on circumstance. The time period between initial occupancy of the facility by the animals and renovation or destruction depends on how well the FAMT performed in each of the five previous design phases. However, once the facility becomes operational, extensive remodeling or demolition should, in theory, never occur because renovation should be an ongoing process.

CONCLUSION

The effective management of naturalistic animal environments requires the formation of an interdisciplinary facility and animal management team whose administrative philosophy must interface with the rationales for developing the facility. The administrative philosophy adopted by the FAMT must also provide the framework from which there will be continuity in the management decision process within and between the six phases of facility design. The range of scientific and technical expertise provided by the FAMT will lead to a comprehensive understanding of the impact management procedures will have on each of the phases of facility design.

It is also imperative that decisions made by the facility and animal management team be naturalistic in both concept and application. A naturalistic approach to facility and animal management involves substantially more than deciding how much food and water should be provided to the animals, and when to clean the facility. It requires knowledge of the physical features of the species natural habitat that are essential for survival and the normal expression of behavior (see Glickman and Caldwell, chap. 14, this volume). A naturalistic approach to management involves the incorporation of these physical features into the captive environment in a manner that will ensure the health and well-being of the animals. As Hediger (1969, p. 20) has stated: "the ideal solution . . . is not to provide an exact imitation of the natural habitat, but rather to transpose the natural conditions in the wild, bearing in mind biological principles, into the artificial ones of [captivity].

As a result of transposing key elements of an animal's natural habitat into the artificial conditions of captivity, an ecosystem is established within the captive environment. Like an ecosystem in nature, the man-made ecosystem in captivity is a sensitive balance between each of the physical features of the environment, and their effects on the animals. Thus, the FAMT must consider the effect of management decisions on the web of ecological interactions that exist within naturalistic environments. The regular monitoring and evaluation of the impact of management procedures on the ecological matrix within naturalistic facilities will help to ensure that physical features of the environment will continue to meet the original design standards. It is only from an interdisciplinary quantitative assessment of management procedures that there can be any assurance for the long-term health and well-being of animals in naturalistic environments in captivity.

MICHAEL K. STOSKOPF
EDWARD F. GIBBONS, JR.

11

Quantitative Evaluation of the Effects of Environmental Parameters on the Physiology, Behavior, and Health of Animals in Naturalistic Captive Environments

A major issue in the design and assessment of naturalistic captive environments is the need to evaluate the effects of environmental parameters on the health and well-being of the resident animals. While this need is intuitively obvious, it is not methodologically simple to obtain reliable and valid environmental, physiological, and behavioral data for the evaluation of animals' health status. It is, therefore, necessary to define precisely what information is to be collected, and what methodologies are to be used to gather, process, and analyze the data. It is only from the application of rigorous research methods that integrative environmental, physiological, and behavioral assessments can be made as to the health and well-being of animals in naturalistic captive environments.

MONITORING PHYSICAL ENVIRONMENTAL PARAMETERS

It may not always be obvious which physical environmental parameters may affect the health and well-being of animals in naturalistic captive environments (Besch and Kollias, chap. 7, this volume). It may also be difficult to determine how these parameters will interact with the animals' physiology and behavior to influence health. Even within the narrow confines of the minimum and maximum tolerable levels of a given environmental parameter for a given species, there can occur major effects on physiology and behavior (Stoskopf, 1983).

The selection of environmental parameters to monitor should be based, in part, on the species' evolutionary history, natural ecology, and life history. For example, if the animals being studied have evolved to occupy a very narrow niche relative to a physical parameter (e.g., as might occur with island endemics), it is reasonable to assume that fluctuations or variations in that parameter might have a significant impact on the health and well-being of the animals as evidenced by changes in physiologic or behavioral function. On the other hand, animals that have evolved to live in environments (e.g., temperate regions) with widely fluctuating values of the same parameter may not be as adversely impacted by modest changes, and monitoring that parameter may not yield useful data. It is also important to keep in mind that the ability of animals to adapt to fluctuations of an environmental parameter may change over time. Dramatic illustrations of this are the adaptations of bear and rodent physiologies to environmental temperature and other factors in relation to hibernation (Canguilhem, 1987).

Time Frames and Data Collection Intervals

Temporal considerations in the monitoring of physical environmental parameters usually involve diurnal, seasonal, or annual cycles. The temporal frame used in a study will affect the frequency with which data are collected. For example, in studies interested in the seasonal impact of temperature, it may be acceptable to monitor this parameter on a daily or even a weekly basis, while standardizing the data collection procedure for time of day and location of measurement. Generally, these studies are long-term investigations, taking an entire year for each iteration of the temporal cycle. This sampling strategy will reduce the hours required to collect information from manually operated monitoring stations. However, the establishment of a sample size that is adequate for statistical assessment will require multiple years of data collection. A more intensive collection of data within a seasonal period may not add to the understanding of the relationships being studied, and will significantly increase the time and effort required to process and analyze the information. On the other hand, if the investigation requires the documentation of physiological or behavioral responses to shorter-term changes in an environmental parameter, infrequent sampling could entirely miss the relationship between that parameter and the response.

Sampling methodologies can also be designed to focus on specific characteristics of a parameter's cycle, which may provide the necessary baseline information to formulate more intensive data collection strategies (cf., Croxton and Cowden, 1946). In certain instances, it may be necessary to evaluate animals' physiological and behavioral responses to extreme environmental values. In such cases, emphasis will be placed on gathering information on the inflection points (i.e., highest and lowest) of a parameter's cycle. This sampling procedure, however, does not provide information about each parameter's duration or rate of change, which may be important to the assessment of animals' physiological and behavioral function. For example, a rapidly dropping (or rising) temperature may elicit entirely different physiological and behavioral responses than gradually decreasing (or increasing) temperatures, even when the same minimum and maximum temperatures are reached within a cycle (Kenshalo, 1990; Swan, 1974).

Automated systems for monitoring environmental parameters have greatly reduced the labor required to collect large numbers of data points at very short intervals over long cycles (cf., Fritschen and Gay, 1979). When such equipment is available there is a tendency to collect as much data as possible within a sampling period. The advantage of such a data collection strategy is that it may permit the post-hoc testing of hypotheses not considered in the original methodological design. The disadvantages are that the collected data may not be qualitatively and quantitatively sufficient to thoroughly examine the original hypotheses. Further, there will be an increase in the probability of making errors in the processing and analysis of data, and there will be increased technical labor costs associated with the administration of the study. In a well-designed study there is a compromise between collecting not enough data and too much data (cf., Croxton and Crowden, 1946; Hays, 1973; Martin and Bateson, 1986). The effective synthesis of what is known about the animals' physiologic and behavioral responses to environmental parameters, and past experiences working with such data, will help to define the proper sample size to effectively evaluate animals' health and well-being.

There are numerous parameters of the physical environment that could be studied in relation to the health and well-being of animals in naturalistic captive environments. However, within the scope of this chapter it is necessary to confine comments to those few parameters that are frequently evaluated when studying the impact of environmental conditions on the animals' physiology and behavior: temperature, humidity, light, and sound or noise.

Temperature and Humidity

Temperature and humidity are perhaps the most commonly monitored environmental parameters, because the technology used to gather information is relatively simple and available at low cost. An important question in designing a monitoring system for these parameters is what substrate is most important to the animals? Air temperatures, for example, may not be as important to the health of burrowing animals as is ground substrate temperatures (Davenport, 1985). Water temperatures may be most important to the health and well-being of aquatic or semiaquatic species (Wilber, 1964). Of additional importance is the requirement that recordings of temperature and humidity (or for that matter any parameter of the physical environment) accurately reflect the range of values that may exist within each targeted substrate. If, for example, the issue of interest is occurring within a den or burrow, collection of data from equipment mounted at eye level on a post in the open air may be of little value. The same is true of questions involving animals that use large ranges in vertical height. Variations in temperature or humidity between the lowest and highest points of the animals' vertical space may be of significance to the study.

The classic tool used to monitor temperature is liquid contained in a glass thermometer. Differential expansion between the liquid and the glass allows measurement of the environmental temperature (Middleton, 1966). The least expensive automated collection systems are bimetallic thermometers. In these, differential expansion of two different metals is used to mechanically move a pointer or stylus. More modern temperature monitoring equipment employs resistance thermometers that function by measuring the change in resistance of conductors or semiconductors. Most of these devices use a null balance system, where the resistance change is balanced and measured by adjusting at least one other resistance in the bridges. The sensitive elements in these thermometers are often platinum, nickel, or copper. Also available are thermistors, which are solid semiconductors consisting of ceramic materials with high-temperature coefficients of resistance. Thermistors are more sensitive to changes in temperature than other monitoring devices, and provide accurate readings of small changes in temperature (Fritschen and Gay, 1979; Sapoff, 1972).

Humidity is the water content of a substrate. In standard usage it refers to the water vapor content of the air, but it can refer to water content of soil or other substrates (e.g., Taylor and Kijne, 1965). Humidity is expressed in several measures: relative, absolute, mixing ratio, and specific (Harrison, 1965). Relative humidity is the ratio, in percent, of the moisture in the substrate (air) to the moisture it would hold if it were saturated at the same temperature and pressure. This is a good index of dryness or dampness, and relates well to evaporation or absorption rates. Absolute humidity is the weight of water vapor in a unit volume of substrate, and is more commonly employed when studying solid substrates. The humidity mixing ratio refers to the weight of water vapor mixed in a unit mass of dry substrate. Specific humidity is the weight of water per unit mass of moist substrate. Humidity mixing ratio and specific humidity are nearly the same, and are less frequently employed in studies.

There are three main types of hygrometers: mechanical, electrical, and dew-point (cf., Ruskin, 1965; Wexler, 1970). Mechanical hygrometers sense moisture through the expansion and contraction of an organic material, often hair held in tension (Davey, 1965). Electrical hygrometers measure the change in resistance of a hygroscopic substance. A dew-point apparatus or psychrometer measures the relative humidity through its relationship to temperature. A psychrometer is one of the oldest and perhaps the most frequently employed instrument for measuring the water content of the air (Fritschen and Gay, 1979). It actually consists of two thermometers, with the sensing area of one covered with a wet cloth or a thin film of water, while the other sensing area is kept dry. The temperature on the wet sensing area will be lowered by evaporation, thereby giving a relationship in equilibrium that allows the calculation of the water vapor content of the air.

Light

Next to temperature and humidity, light may be the most studied environmental parameter in relation to the health and well-being of animals in captivity. From a medical perspective, the installation of artificial lighting systems in naturalistic facilities may be preferred over natural lighting, because they will permit the control and regulation of four components of light that are cited as important to the health and well-being of animals: energy, intensity, spectrum, and periodicity (Bellhorn, 1980; McSheehy, 1983; Weihe, 1976b). The use of direct sunlight in naturalistic captive environments can provide too much heat energy, and in a relatively short time may thermally load small enclosed habitats. Animals from latitudes significantly different from their holding facility may, in such instances, be subjected to inappropriate light intensities, spectrum, and periodicity.

Studies are needed that determine how much variation in light intensity from natural light conditions is allowable for a given species. Such data are of importance, because lux are not necessarily cumulative, and the strategies of delivering either less intense light for longer periods of time each day or more intense light for shorter periods of time do not necessarily compensate for inappropriate light intensities (O'Steen et al., 1972).

The intensity of light can be measured in many ways, but most appropriate for clinical purposes is the amount of light delivered to a photographic white reflection board that can be moved to specific locations within naturalistic captive environments. The reflected light is measured in lux which is a unit of intensity related to the foot candle (Keitz, 1971). Expensive lux meters are available, however in situations where intermittent readings are required, a photographic light meter or a single lens reflex camera with through-the-lens metering can be used to obtain lux readings (Table 11.1).

Spectral data are more difficult to reliably obtain. Equipment for measuring spectra is expensive and although the spectral output of a bulb is usually discussed in terms of wavelengths of light emitted, a convenient way of characterizing the light spectrum is through color temperature in degrees Kelvin (Stimson, 1974). A high color temperature, one with high degrees Kelvin, has a lot of blue in its spectrum.

The importance of providing staged illumination varies with the species being maintained, but it is difficult to go wrong with a well-designed crepuscular period. This is facilitated in systems where several light sources are used to achieve adequate lux and appropriate spectrum. Measurements of periodicity are easily measured with the use of photovoltanic cells that activate a recording device or transmitter.

Sound

Sound plays an important role in the lives of most species of animals. The potential impact of sound on the health and well-being of animals should not be overlooked (Fletcher, 1976; Stoskopf, 1983). Sound, like light, has four components that may affect animals' health either independently or in concert: intensity, frequency, periodicity, and content.

TABLE 11.1
Converting Photographic Light Meter Readings to lux
(Camera or Meter Sct to ASA 50 or 18 DIN)

LUX	f Stop	Shutter Speed	
2.8	2.0	19	
2.8	1.0	38	
2.8	0.5	75	low light plants
2.8	0.25	150	
2.8	1/8	300	
2.8	1/15	600	
2.8	1/30	1200	
2.8	1/60	2400	
2.8	1/125	4800	
2.8	1/250	9500	plants in nature
4.0	1.250	19000	
5.6	1/250	38000	
8.0	1/250	75000	bright daylight sun

From Stoskopf, 1992

The effects of high intensity sound on animals' health have been documented (Jurtshuk et al., 1959; Rai et al., 1981). Sound intensity is measured in decibels with a decibelmeter. A decibel is a logarithmic unit that measures the change in intensity of sound. A neper is a similar unit based on natural logs. One neper equals 8.686 decibels. Both units are proportional to the square of the sound pressure. Sound pressure refers to the incremental variation in static pressure of a medium when a sound wave is propagated through it. It is measured in microbars. One microbar is approximately one-millionth the normal atmospheric pressure (Goldstein, 1978).

Although sound intensity is most frequently evaluated in relationship to health and well-being, its effects may not be independent of other parameters. Certainly, acoustical interference can have a dramatic effect on the behavior of captive animals (Clough, 1982). Motors or other environmental noises (including those made by observers) may completely disrupt baseline behavioral activity. The temporal aspect of sound periodicity is also a critical issue, and may determine the effect of a sound on an animals' physiological state. Continuous sounds of high intensities may cause more severe changes in biochemical baselines than intermittent sounds of similar intensities (Ising, 1981). Acclimation of animals to regularly repeating sounds may have different effects than random sound events. This question requires further investigation.

The frequency of sound may have significant effects on animals' physiologic state (Evans and Tempest, 1972; Broner, 1978). Studies on the impact of low frequency noise, below the hearing threshold, on the neuroendocrine axis of humans (Evans and Tempest, 1972) may well indicate a similar impact on other species. The advent of low cost digitizing and frequency analysis software for personal computers has made investigations into these areas much more accessible than in the past. Expensive equipment can now be replaced with a broad frequency microphone and a MIDI connection into a computer.

The final aspect of sound that can certainly affect the physiology of humans and most probably has similar affects on other species, is content. Sound content, for example, can play an important role in the control and modification of animals' behavior, including pests (Frings and Frings, 1968; McConnell, 1991). Content, however, is the most difficult aspect of sound to evaluate.

Without careful baseline studies devoted to the socioecological meaning of sound content, it will be difficult to evaluate the impact of sound content on the physiology and behavior of animals.

MONITORING PHYSIOLOGICAL PARAMETERS

It is imperative that the physiological parameters examined in any investigation concerning the interrelationship of environment, behavior, and health be carefully selected and relevant to the questions being asked. In questions related directly to health, some presumptions as to the probable health impact of environmental factors must be made to focus the collection of physiological data. In mammals, an environmental factor can affect one or multiple systems or organs. The predicted pattern of systemic or local organ impact will determine to a great extent which physiological parameters are most affected.

Most commonly one or more of three major physiological systems are monitored: the immune system, the reproductive system, or the xenobiotic metabolizing system. All three systems play important roles in the health of animals, and are affected by environmental and psychological factors (Follett et al.,1985; Gut et al., 1981; Kelley, 1985). The functioning of all three systems can be evaluated through either remote or interventional monitoring.

Immune System

The immune system provides the major defenses against infectious disease. It consists of two major components: cellular and humeral defenses. Both components of the immune system can be modified by environmental factors through endocrine action on bone marrow, thymus, or circulating cells (Blalock, 1984; Kelley, 1985; Voller and Bidwell, 1988). Increased production of corticosteroids affects activated B lymphocytes, reducing antibody production, and making animals more susceptible to infectious disease. Efforts to monitor this aspect of physiology are often indirect, and look for the effects of impending infectious disease (Werner, 1983). Hematological examinations, including total and specific blood cell counts, are good examples of this type of monitoring.

Reproductive System

Evaluation of the reproductive system may center on reproductive success measured in the ability of animals to sire and to rear viable offspring, which can be monitored with infrequent physiological evaluations (Beck and Power, 1988; Pevet, 1987; Lasley, 1985). On the other hand, specific questions regarding spermatogenesis, ovulation, menstrual cycle regularity in primates, or reproductive hormone levels (all end points that can be affected by environmental or behavioral factors) may require more significant intervention to acquire adequate samples for analyses. Reproductive hormonal levels can be measured in samples of plasma, serum, urine, and feces through the use of radioimmunoassays. There are, however, methodological and species-specific problems associated with the use of radioimmunoassays that may affect an assay's validity: specificity, accuracy, precision, and sensitivity. Discussion of such considerations can be found in Midgley et al. (1969), and Reimers and Lamb (1991).

Xenobiotic Metabolizing System

The metabolism of substances that are not normally found in a biological system is termed xenobiotic metabolism. The primary goal of xenobiotic metabolism is to make foreign compounds more water soluble to facilitate excretion. This is done through a number of enzymes, many of which use cofactors in their catalysis. The availability of these cofactors, and the actual enzymes involved in the catalysis itself, can be limited by depletion of intracellular ion concentrations or other effects of the neuroendocrine axis (Dauterman, 1980; Hodgson, 1980). Alterations in the net xenobiotic metabolizing ability of animals will alter the effects of foreign substances and drugs on the specimen. This can have significant impact on the behavior of animals in captivity. For example, immobilization times or responses to anesthetics may be significantly altered (Eriksen, 1978; Harthoorn, 1976). It will also change the pattern of excretion of xenobiotics being ingested or absorbed (Doull, 1980). Quantitative analyses of urine and feces for metabolic end-products of xenobiotic compounds can provide information on the state of the neuroendocrine axis.

Remote Monitoring of Physiological Systems

The ideal situation when monitoring physiological parameters in animals is the use of remote monitoring supplied by an implanted or attached telemetry device (cf., Amlander and MacDonald, 1980; Cheeseman and Mitson, 1982; White and Garrott, 1990). The science of this form of physiological measurement is developing rapidly. However, when considering the use of biotelemetry to monitor physiological parameters in naturalistic captive environ-ments, it is important to determine if there will be any interfer-ence in the reception of the transmitted signal as a result of elec-trical noise and commercial transmission signals (Smith, 1980).

Among the various biotelemetry devices that can be used to monitor physiological parameters are internal temperature trans-ducers that can either be implanted in the abdomen or swallowed, and can be used to detect pyrexia or fever on a real-time basis. They can also be used to detect changes in body temperature that are associated with ovulation or parturition. Implanted sensors have the advantage of longer reporting life, because swallowed transducers are usually limited to the gastrointestinal transit time of the animal being studied. On the other hand, surgical implanta-tion generally requires considerable time for the subject to return to baseline physiological and behavioral states. The usefulness of any implanted or swallowed sensor depends heavily on its biocom-patibility (Reite, 1985), and the range and ease with which the data can be detected and reconstructed into a data set (Cunningham and Peris, 1983; Kroll and Johnson, 1980; Varosi et al., 1990).

The remote measurement of animals' body temperatures can also be achieved through the use of thermography and two-color pyrometers. Thermography measures infrared radiations emanat-ing from an animal (or a substrate). This technology uses heat sensing crystals, indium antimony, or mercury-cadmium-teluride to detect the radiation and to calculate temperatures (Reed, 1972). Two-color pyrometers measure the ratio of two wavelength regions to determine temperature (Warnke, 1972). Either of these devices provides data on surface temperatures of an animal as opposed to core body temperatures.

Other remote sensor systems used in monitoring the physio-logical condition of animal subjects include electrocardiogram and heart rate monitors (Johnson et al., 1980; Stohr, 1988; Walcott, 1980). Recently, passive computer chips have been designed that can be placed in line with the vascular system to provide real-time data on blood pH, oxygen content, and carbon dioxide levels. Unfortunately, the range of these devices is very small and imprac-tical for studies conducted on free-ranging animals. Future devel-opments are expected in the real-time telemetry of specific blood electrolytes.

A less sophisticated approach to remote monitoring is the retrieval of excreted body wastes. Urine and feces can provide con-siderable information if properly collected and analyzed (Reimers and Lamb, 1991). Questions regarding the possible contamination and reliability of samples are best answered through multiple analyses involving compounds with different sources and stabilities. For example, reproductive hormone metabolite assays are frequently standardized against creatinine in the urine (Taussky, 1954). Modification of ritual urination or defecation sites may greatly facilitate sample collection.

Interventional Monitoring of Physiological Systems

When remote monitoring is not practical or available, it is necessary to perform interventional health monitoring, where the animals being studied are manipulated and subjected to physical examination and sampling (Balk, 1983). Frequently, hematological and serological monitoring are used to obtain an assessment of each subject's overall health. Hematological values and many serum enzymes change predictably with disease and can be used to indirectly evaluate the proficiency of the immune system (Outteridge, 1985; Vyas et al., 1975). Serum can be directly analyzed for hormones of the neuroendocrine axis, xenobiotic compounds or their metabolites, or reproductive hormones. Serum or plasma samples can be collected relatively quickly and easily, and there is usually no long-term debilitating impact on the subjects other than what may occur as a result of chemical or physical restraint of the animals. Other interventional samples that can be collected include biopsies and samples of other body fluids, but these are much less frequently employed.

When employing interventional techniques it is important to standardize as much as possible the procedures for obtaining data. Variation in the time periods and manner of interventional manipulation should be kept as minimal as possible throughout the study. Such methodological continuity will minimize intersample variations due to effects of immobilization or restraint on the animals.

MONITORING BEHAVIORAL PARAMETERS

Technical and practical difficulties in the physiological evaluation of animals' health status necessitates monitoring behavior. The systematic collection of behavioral data augments findings from physiological tests, and may serve as an indicator of impending health problems. Behavioral data can also help to define environmental parameters that affect the health status and well-being of animals (Gibbons and Stoskopf, 1989).

For behavioral data to be a valid instrument in health assessment, it is imperative that rigorous research methods be used to collect, process, and analyze the data. These methods include the definition of behaviors specific to the evaluation of health status, the training and periodic monitoring of research personnel, the establishment of a data base large enough to reliably define behavioral health parameters for the target species and respective animals, the reliable coding and processing of data, and the appropriate application of descriptive and inferential statistics. These and other methodological considerations have been addressed in a number of publications. The reader is referred to this literature (e.g., Altmann, 1974; Banks, 1982; Colgan, 1978; Dixon and Chapman, 1980; Denenberg and Banks, 1969; Dunbar, 1976; Hazlett, 1977; Hinde, 1973; Hutt and Hutt, 1970; Lehner, 1979; Martin and Bateson, 1986; Moberg, 1987; Pearl and Schulman, 1983; Sackett, 1978; Slater, 1973; Tinbergen, 1963). The following discussion will focus on select methodological issues as they pertain to the assessment of animal health status in naturalistic captive environments.

Defining Behaviors

Defining units of behavior is fundamental to any systematic study involving the relationship between animal behavior, health, and well-being. The defining of behavior may include a complete description of species-typical behaviors, or may characterize a subset of behaviors that are important to assessing health problems (e.g., Altmann, 1965; Martin and Bateson, 1986). Failure to pay careful attention to the definition of behavioral units will increase the likelihood of errors associated with the observation, coding, and processing of data and will affect the validity of the study (Lehner, 1979).

When defining behaviors for the evaluation of animal health status in naturalistic captive environments, a few general principles should be kept in mind. First, the spatial complexity of naturalistic captive environments may make difficult the observation of subtle variations in behavior. Second, not every change in behavior indicates a deterioration of health status and well-being. Behavioral changes over time may be a function of animals' adapting to aspects of their environment (Gonyou, 1986). Third, behavioral data must be quickly processed and evaluated. The medical team cannot wait weeks (or in some cases even days) for analyses to be completed. Therefore, the initial focus of such investigations should be on general aspects of behavior, such as percent time spent active versus inactive, appetitive and positional behaviors, spatial usage of the environment, and occurrences of stereotypes. Significant changes in the frequency of occurrence or form of these behaviors should alert the observers that "something may be wrong," and more specific behavioral observations and physiological tests conducted to closely examine the animals. As a further note, behavior alone cannot define a change in animals' health status and well-being (or for that matter, any physiological state). An interdisciplinary approach must be used, and should include: physiological findings, behavior, the physical appearance of the animals, and caretaker daily log books (e.g., Gibbons and Stoskopf, 1989).

Sample Size

The amount of behavioral information necessary to detect changes in health depends largely on the number of independent variables defined as important to the assessment. Other factors that influence sample size include the sampling procedure used to collect the data and the desired power of the statistical tests. In general, the larger the number of independent variables and the greater the desired statistical power, the larger the required sample size (cf., Cohen, 1977; Hays, 1973; Lehner, 1979; Martin and Bateson, 1986; Sokal and Rohlf, 1981).

Having an adequate sample size is not a trivial issue. It has methodological importance when the collection of baseline (i.e., normative) data is interrupted by animal illness or other husbandry problems. In such instances, the size (N) of the baseline data may be reduced by the animals' medical treatment or removal from the environment. Any decline in animal health may affect the validity of normative data (Weihe, 1988). Specifically, because of physiological changes in the animals not accounted for in the research methods, the data base may not reflect true qualitative and quantitative aspects of normative behavior for the species and the individual animals. Further, it may be impossible to detect at what point in the data collection process the physiological changes first occurred, thereby compromising the entire data set. Concerns regarding proper sample size and validity of baseline and subsequent data should be given highest priority in the behavioral assessment of animal health and well-being.

Normative Behavior

Knowledge of the range and frequency of species-typical behaviors is fundamental in the use of behavior as a health assessment tool. It is important to consider not only those behaviors that occur with regularity, but also those behaviors that occur under specific physiological or environmental conditions (see Glickman and Caldwell, chap. 14, this volume). The total range of behaviors that a species may employ can be defined as its behavioral totipotentiality, and may be estimated from analyses involving the interface between morphology, physiology, behavior, and environment (Jouffroy and Lessertisseur, 1979; Prost, 1965). Studies conducted in the wild and in captivity can be important sources of information for the determination of the range and types of behaviors expressed by a species. Complementary field and captive studies may also provide insights into the proximate and ultimate influences on behavior (Janson, chap. 18, this volume; Tunnell, 1977).

Normative behavioral profiles should also be established for individual animals. This requires knowledge of each animal's medical, demographic, environmental, and rearing histories (Sanford et al., 1986). In general, the range of behaviors used by individual animals (i.e., habits) will be a subset of the species' behavioral totipotentiality, and different animals may engage in distinct subsets of behavior (e.g., Prost, 1965). The behavioral habits of individual animals may change over time as a function of endogenous and exogenous events. It is the interanimal and intraanimal variability in the expression of behavior that presents the greatest obstacle to the use of animal behavior in the assessment of health status and well-being (Moberg, 1987).

Statistical Strategies—Correlation and Causation

A common error made by animal managers not well versed in research methods and statistics is to conclude a cause-effect relationship between environmental events and declines in animals' health based solely on naturalistic studies of behavior. In naturalistic studies, the data are best suited for correlational analyses that measure the degree of covariation between environmental events and animals' behaviors. Correlational analyses do not define the reasons for the covariation. It is only through the direct manipulation of independent variables that cause-effect relationships can be concluded between environmental events and behaviors that may indicate a decline in animals' health status and well-being.

The inability to conclude cause-effect relationships between environmental events and health status does not rule out the importance of naturalistic observations of behavior. It is only from behavioral studies that the health management team can begin testing cause-effect relationships through the use of controlled experiments. The effective use of naturalistic observations and controlled experimentation is an important adjunct to any health assessment program.

Specific Behavioral Examples

Attempts to identify specific behavioral patterns that will reliably define deterioration in animal health status have met with mixed results. A number of behaviors have been found to be suggestive of changes in animal health and well-being, but none can be cited as definitive evidence of medical problems. The primary reason for the lack of direct association between behavior and health is the interanimal and intraanimal variation in behavioral response to environmental events that may cause medical problems (Moberg, 1987). Unless the medical team can determine the motivational mechanisms and roles of behaviors in animals' lives, it will be difficult to associate the occurrence of a particular behavior with disease (Broom, 1987).

Consequently, the medical management team should look for constellations of behavior that may signify a decline in health status. Groups of behavior that may be used include: sterotypies, restricted use of spatial environment, changes in social interactions, emittance of particular vocalizations, abnormal forms of locomotion and posture, behavioral lethargy or excessive activity, changes in appetitive behaviors, and lack of grooming behavior as indicated by matted fur and a general unkept appearance (Colam-Ainsworth et al., 1989; Dantzer, 1986; Duffell et al., 1986; Ewbank, 1985; Morton and Griffiths; 1985; Nicholls and Handson, 1983; Sojka, 1986; Soma, 1987; Webster, 1983). The occurrence of one of these behaviors should not be construed as definitive evidence of a change in health status. If, however, observations indicate the occurrence of a number of these behaviors during a specified time period, the medical team should be alerted to the possible existence of a health problem.

ENVIRONMENTAL, PHYSIOLOGICAL, AND BEHAVIORAL
MONITORING: INTERFACE BETWEEN DATA SETS

The development of a comprehensive profile of animals'
health status and well-being requires the quantitative integration
of environmental, physiological, and behavioral data sets. Careful
attention to the methods and statistics used to develop and compare data sets will provide measures that yield valid health assessments (Dubin and Zietz, 1991). Considerations that may affect the
validity of comparisons between environmental, physiological, and
behavioral data sets include the time schedules used to collect
individual data points, possible interference in collection methods
between data sets (e.g., the impact of physiological recording on
behavioral data collection), and the statistical measures and decision rules used in the assessment of animals' health and well-
being.

Temporal Synchronization of Data Sets

Temporal synchrony in the collection of environmental, physiological, and behavioral data will help to ensure valid comparisons
between data sets. Data for each set can be collected either continuously or at different points in time. The decision as to which
temporal strategy to use will depend on technical aspects of data
collection (e.g., equipment used, and compatibility of methods
between data sets), level of observational detail required for the
collection of each data point, and the time and expense associated
with the collection and processing of information within and
between data sets.

Continuous data collection is to be used when an exact
record of an event is required. Continuous recording will yield
exact information regarding each event's trend over time, including frequency and duration of occurrence. In most cases, however,
continuous data collection will limit the number of events that
can be monitored within and between data sets (Martin and Bateson, 1986). In naturalistic captive environments, the continuous
collection of environmental data is recommended when information can be recorded using remote sensors and stored directly on
computer, thereby minimizing interference with the animals'
physiological and behavioral functioning and the collection of
respective information. The continuous recording of physiological
or behavioral data, however, can be technically difficult and pro-

hibitively expensive, even when information is recorded using biotelemetry.

Time sampling strategies, on the other hand, will not provide an exact record for any given event, but will permit the reliable observation of a number of events (Lehner, 1979; Martin and Bateson, 1986). Of consideration in the use of time sampling is the temporal interval between observations. If, for example, the interval is only a few seconds, then many of the benefits and costs associated with continuous sampling will come into play. An interval of longer duration (e.g., one or two minutes) will permit the recording of many events either simultaneously or offset by a standardized time. The decision as to what time interval to use should be based on the number of events to be monitored, the equipment used to monitor each event, and past experiences working with such data sets. A few test trials will help to determine the best sampling interval. It should also be kept in mind that regardless of the length of the sampling interval, the collection of each data point should be as temporally discrete as possible (Martin and Bateson, 1986). In naturalistic captive environments, the recording of physiological and behavioral data using time sampling may prove to be technically and economically cost-effective especially when interventional methods are used to record animals' physiological functioning.

The objective to minimize methodological and technical interference between data sets will be critical in the decision to record data either continuously or by time sampling. There are an unlimited number of ways that interference can occur between the methods used to collect environmental, physiological, and behavioral data. Therefore, it is important to decide which combination of methods and instrumentation are most compatible and will provide a valid integrative assessment of animals' health status and well-being. For example, the collection of environmental, physiological, and behavioral data at different points in time will help to minimize methodological interference between each of the data sets. However, such a sampling strategy may make difficult comparisons and interpretations of findings between data sets.

Statistical Considerations and Decision Rules

Huff (1954) in his classic book, *How to Lie with Statistics*, humorously illustrated how data can be statistically manipulated to support almost any hypothesis. The lesson one learns from this volume is that any reasonable arithmetic operation can be applied to numbers with the result producing mathematically valid numerical values. The problem arises, however, when meaning is applied to the resultant numerical values. If the mathematical operations conducted on data do not follow some rule of logic inherent in the purpose for compiling the information, then it would be impossible to say anything meaningful about the results. In other words, nonsense in = nonsense out.

The same guiding principle holds true when using statistics to arrive at conclusions regarding the health and well-being of animals. If there is no guiding methodological philosophy to orient data analysis, then there will be little hope of extracting anything meaningful and valid from the data. A methodological philosophy for data analysis begins when the research question is defined. Development and refinement of the philosophy continues when the subjects are selected for study, and there is a formal research design, including decision rules (e.g., alpha levels). As a direct result of a given philosophy, the information recorded will be assigned numbers according to specific levels or scales of measurement: nominal, ordinal, interval, or ratio (Stevens, 1946, 1951).

The level of measurement for any set of data will determine which statistics may be used, and what conclusions may be drawn from the findings. Some controversy exists as to which statistical procedures may be applied to a given level of measurement (cf., Hays, 1973, pages 81–90). However, it is suggested that the following rule of thumb be applied: if the data is of nominal or ordinal scales, then use nonparametric statistics. If the data is of interval or ratio scales, then either nonparametric or parametric statistics may be used in the analysis (Lehner, 1979). It may also be of value to note that nonparametric measures, when properly applied to data, can be as statistically efficient as comparable parametric tests (Bradley, 1968).

The issue of levels of measurement, statistical procedures, and conclusions becomes substantially complicated when analyses involve data sets of different levels of measurement, as may occur when evaluating environmental, physiological, and behavioral data. There is no formal statistical rule that dictates what to do when faced with this situation. The best advice that can be given is to use statistics that apply to the data set with the lowest scale of measurement. For example, if the recorded environmental and physiological data were interval in scale, while the behavioral data set was ordinal in scale, then it might be prudent to use statistics applicable for ordinal scale data when comparing the data sets.

It is also important to recognize the interrelationships between animals' physiology, behavior, and environment. It is because of this interface that the physiological and behavioral functioning of animals should not be considered as independent events. This fact must be taken into account when applying statistical measures to physiological and behavioral data, and in the interpretation of the results (Martin and Bateson, 1986). Failure to recognize that physiological and behavioral data sets are not independent, may artificially increase alpha (α, the probability of making a Type I error), which may lead to the rejection of the null hypothesis when it is actually true. In other words, without knowing the actual increased value of α, it might be decided that the animals are experiencing declines in health and well-being when actually there is no reason for medical intervention.

There are situations, however, when making a Type I error may offer advantages in the evaluation of animals' health and well- being. When the probability of making a Type I error can be directly controlled, it may be a powerful tool in preventive medicine. This type of decision is referred to as a "false positive," and is the α level used to reject the null hypothesis. In most scientific studies α is set at 0.05. However, when making decisions regarding the health and well-being of animals, it may be useful to set α at a slightly higher level (e.g., 0.07, 0.10). This health decision strategy is particularly recommended when rejection of the null hypothesis results in acute monitoring of the animals, not direct medical intervention (e.g., capture and hospitalization). By being aware of and directly controlling the probability of making a Type I error, decision strategies can be made more sensitive to physiological and behavioral changes that might signify the onset of health problems. The early detection of changes in animals' physiologic and behavioral functioning will minimize costs associated with the medical care of the animals, and will help to ensure the health and well-being of animals in naturalistic captive environments.

CONCLUSION

The evaluation of the physical environmental features of naturalistic captive environments should begin when animals are chosen for the facility, and continued through the design and construction phases of these environments (see Doherty and Gibbons, chap. 10, this volume). Critical to the evaluation process is the formation of a health management team whose primary responsibility will be to establish a comprehensive assessment protocol that will effectively evaluate and manage the health and well-being of the animals. Because the assessment of the health status of animals involves all phases of facility design, construction, and operation, the health management team must consist of professionals from a variety of disciplines, including animal caretakers.

The incorporation of the health management team into the development of the facility should lead to the identification of design and construction flaws that might have a negative impact on the health and well-being of the animals as early as possible in the construction phase. The evaluation of animals' health and well-being should be a continuous process, and information from a variety of disciplines should be collected throughout the development and operation of the facility. It is imperative that data be gathered on relevant physical environmental parameters, and on the animals' physiological and behavioral functioning. It is also crucial that rigorous attention be paid to the methods and equipment used to collect the data, and the statistics employed to analyze the information. The difference between casual data collection, and obtaining environmental, physiological, and behavioral data that are scientifically valid can dramatically influence the decisions of the health management team regarding the health and well-being of the animals. Conversely, incorrect evaluations of the health status of captive animals can compromise the validity of any research conducted on the animals, thereby diminishing the important value of naturalistic captive environments for behavioral research.

IV. THE IMPORTANCE OF
      NATURALISTIC ENVIRONMENTS:
      RESEARCH ON SELECTED SPECIES

MEREDITH WEST
ANDREW KING

12

Research Habits and Research Habitats: Better Behavior through Social Chemistry

If bird watchers or ornithologists had any say in the matter, the topic of habitats for cowbirds would be an open and shut case. Cowbirds would be jailbirds. Branded the "black vagabond" (*Molothrus ater*) in avian taxonomy, cowbirds evoke striking attributions of criminality from humans: "The cowbird is an accomplished villain, and has no standing in the bird world. . . . No self-respecting American bird should be found in his company. . . . As an outcast he makes the best of things, and gathers about him a band of kindred spirits who know no law" (Chapman, 1912, p. 359).

The basic problem is that the evil cowbirds do lives after them, in the form of eggs deposited by females in the nests of over 200 different avian species and subspecies (Friedmann, 1929; Friedmann et al., 1977). Nestling and fledgling cowbirds are then unwittingly but obligingly fed by foster parents, often to the detriment of the "true" brood. It is thus no wonder that the topic of the care and psychological well-being of cowbirds has a very small following.

Our interest in their care relates only indirectly to the species' parasitic habit. By abdicating parental duties, cowbirds would seem to bequeath to their young a host of psychological problems regarding species identity (Roy, 1980). How will cowbirds find other cowbirds? How will they acquire species-typical behavior? Not only are young cowbirds exposed to species-inappropriate behavior from their foster families, but they are also deprived of opportunities to learn from their own kind while still young. Such unusual ontogenetic circumstances led Mayr (1974) to suggest that cowbirds might be an ideal example of a closed developmental program—that is, one in which postnatal experience plays little to no role in transmitting species-typical behavior. It was the nomination of cowbirds as the avian species least likely to learn that awakened our interest in their ontogeny. We set out to look for evidence of different developmental pathways associated with parasitism, with a particular interest in determining what role, if any, learning played in the development of communicative skills (King and West, 1977).

We focused on vocal behavior because learning had been implicated in song ontogeny of all other passerines studied to date (Kroodsma and Baylis, 1982). The discussion here only reveals some of the complexity uncovered thus far about cowbird communication (see also Rothstein et al., 1988). In particular, the discussion focuses on the role of visual and vocal experience in learning to sing effective courtship songs. The plan of the chapter is to review some recent findings on cowbird communication while describing the principles of animal care that govern the work. Dealing with these two goals in an integrated manner is quite appropriate. Our interest in accommodating the social needs of cowbirds led to some of the most unexpected discoveries about their communicative behavior. We begin by reviewing some basic operations of our laboratory.

GOVERNING PRINCIPLES

Experimenter Participation

The first rule in our laboratory is that animals in our care are cared for by us—researcher participation in the daily feeding and maintenance of each animal is considered "quality time," allowing valued glimpses of an animal's everyday existence. The experience can also provoke new research questions. In cowbirds, firsthand knowledge of the multidimensional nature of individual differences led to new ways to analyze song responsiveness (King and West, 1989). For another example of the benefits of participation, readers may want to read about captive starlings (*Sturnus vulgaris*). In this species, the nature of the relationship between human care givers and experimental subjects has striking effects on the sounds starlings mimic, with imitation of human speech occurring only under conditions when humans physically interacted with the birds (West et al., 1983; West and King, 1980).

Social Habitats

The second principle in our laboratory is that animals under our care are housed in social units, sometimes males and females together, sometimes males together, sometimes a male or females with members of other species (see later for an exception). The conditions of solitary housing, routinely used in so many studies of animal learning, are avoided because of risks to "psychological well-being" (Novak and Suomi, 1988). In this regard, hormonal assays of cowbirds indicate that isolation is far from a neutral environment. Male cowbirds housed alone showed elevated levels of plasma B corticosterone relative to free-living males, while captive males housed with females do not (Dufty and Wingfield, 1986). Dufty and Wingfield speculate that the removal of "social stimuli was stressful for these cowbirds" (p. 230). In contrast, captive male cowbirds show levels of testosterone (T) comparable to free-living males, supplying further evidence that the corticosterone response was not a general response to captive conditions. Few data of these sort are available for other species, but it is likely that similar data could be found for comparably social species.

When normally social animals are housed in isolation, the study of development is also hindered because the abnormal conditions lead to neophenotypic behaviors (Kuo, 1966; Johnston, 1988; Lehrman, 1953; Marler and Hamilton, 1966). Isolation does not provide a baseline condition for the study of species-typical development, an assumption made in many studies of songbirds where the behaviors of isolates is equated with identification of "innate" behavior (Marler, 1976). Isolation is more properly termed a condition of "biased rearing," a term that more properly captures the active nature of the experimental manipulation (Gibson, 1969).

Isolation usually does accomplish one unambiguous goal: it separates scientists from their subjects. By virtue of the nature of the housing apparatus, investigators deny themselves access to the most important resource in the study of development—experience of watching continuity and change. They also deny themselves something equally important, the opportunity to see animals "out of work." Isolation can also be considered a sheltered environment in which an animal is freed from responsibilities to find food, shelter or to interact with peers. Animals in nature do not develop species-typical behavior without dealing with the realities of the natural world. How animals learn (or fail to learn) when isolated from conspecifics may not be generalizable to other settings. Not only may imitation or observation of others be more prominent in naturally occurring instances of learning, but even quite basic processes of association or reinforcement might be affected by altering the reinforcing value of conditioned stimuli or responses. So too the motivation to learn may vary with social settings. In the case of cowbirds, attention to songs varies in significant ways depending on the identity of the listeners companion (West and King, 1986; King and West, 1989). We view social housing as a means of altering a species social chemistry to identify the different means by which experiences are compounded into adaptive outcomes.

Social housing carries considerable economic and scientific costs for institutions and investigators. Animal enclosures must of necessity be larger and the number of animals to care typically increases. Investigators must also be able to observe the animals regularly and for long enough durations to determine that the animals treat each other in "humane" ways. The nice thing about housing an animal alone is that it is easy to maintain the scientific fiction that the experiences the animal is undergoing are known. Studying animals two by two, or in any other multiple, leads to a deluge of potentially important information to be described and quantified.

Nested Environments

A third principle guiding our laboratory investigations is the use of different environments, ranging from the confines of a nest to the spaciousness of large multicompartment aviaries to the seclusion of life with one or two individuals in smaller enclosures. Moving birds through these settings simulates what happens to many birds in nature. It is quite natural for them to undergo seasonal shifts in lifestyle that to humans might seem quite radical. Cowbirds can go from living with a brood of foster siblings to small clusters of other juvenile cowbirds to large (on the order of millions) winter roosts to a pair bond with a female before they are one year old. In designing our facilities, we have aimed at making environmental transitions as feasible as possible so that we could witness how such naturally occurring transitions correlate with changes in vocal ontogeny.

Opportunities, Options, and Obligations

A fourth principle requires the balancing of opportunities for observation by humans with opportunities for privacy for the animals, and weighing experimental options against experimenter obligations to ensure animal safety. On the first matter, three to four meter high cages are constructed because cowbirds seem more comfortable looking down on humans. So too, cages have barriers for birds to hide behind and aviaries have different sections, each providing food, water, perches, and trees. The birds can get away from observers, human or avian. This option is vital in studies of dominance or mate choice to simulate natural conditions where animals can decline to participate by departing from field sites.

With respect to the second matter of options versus obligations, the decisions are harder to state in terms of a simple principle. But, in general, the more knowledge investigators gain about the social dynamics of established groups, the less freedom exists to exert experimental control. For example, male cowbirds that are successful in obtaining copulations from females in aviaries sing highly attractive songs as judged by playback techniques described later in this chapter. But there is a question about the direction of effects. Do these males sing potent songs because they are successful breeders (an effect of reproductive status), or are they successful because they have highly attractive songs (a cause of reproductive status)?

One way to disentangle these effects would be to transfer males from one context to another as a function of their mating status or as a function of properties of their song and to test their success with new companions. This manipulation cannot be carried out, however, because the risks of injury from other males are too high. By examining the fates of males transferred from one aviary to another as a function of song properties, we discovered that the males most likely to be injured were those singing highly potent songs (West and King, 1980). Even short-term (four hour) introductions into new settings provoked aggression (West and King, 1980). Dufty and Wingfield (1986) also found that captive males could not establish dominance without singing. Once in possession of this knowledge, an experimental option was "lost" in terms of arbitrary subject assignment to social groups. But the knowledge gained of interrelationships among singing, competing, and courting certainly outweighed the chance to manipulate variables in a statistically clean manner.

Although these governing principles involve more attention to the everyday details of the animal's existence, they afford opportunities for truly ontogenetic studies. In particular, longitudinal work on animals with known life histories and brief intergenerational time periods is possible. In the following sections, we review some of the social and cognitive capacities of cowbirds identified using the housing principles just described.

THE SOCIAL CHEMISTRY OF VOCAL LEARNING

When eastern brown-headed male cowbirds are raised from the egg with no opportunities to hear males sing, but with female cowbirds as companions, they produce highly effective songs. Effectiveness is measured by testing the reflexive reproductive responsiveness of male-deprived female cowbirds to playbacks of the males' songs. Such females adopt copulatory postures to certain cowbird songs: females arch their backs, lower and spread their wings, and separate the feathers around the cloacal area. Not all songs are equally potent. Only those of cowbirds, and typically only those of individuals from the female's natal population, are effective releasers (West et al., 1981). The selectivity of the females renders the procedure even more valuable as it affords a quantifiable index of song effectiveness. Each playback experiment has a six to nine month preparation period, during which time females are housed together out of earshot of males. Testing occurs during only six weeks of the year, the period of the female cowbird's breeding cycle. During that time, females hear six or seven songs a day to keep their threshold low. In nature, they would probably hear hundreds of songs before 8 A.M. The procedure is invaluable in learning about the functional properties of males' songs, but it is expensive in terms of time and resources because so few songs can be tested in a yearly cycle.

The playback procedure also brings to the fore another potential disadvantage of social housing—that is, the possibility of inhibitory or excitatory effects between companions. We house females together prior to testing and during testing, with the exception noted below. When we first published data on female responses to songs, reviewers questioned the interpretability of data from socially housed females. How could we rule out inhibitory or excitatory effects? We used several methods to address the issue. First, we compared patterns of responsiveness between females housed together. Did females housed together always respond or not respond on the same trials? The answer was no. Correlations conducted on playback responsiveness between females housed together were neither highly positively nor negatively correlated. The correlational scores for females living together were also not different from the scores of females living in separate chambers but whose data were paired for the purposes of comparison. Females that had never been in contact with one another had profiles of responsiveness indistinguishable from those of females living together (King and West, 1983a).

We also compared the playback responses of solitary versus pairs of females (thereby violating our social housing rule). All females were exposed to the same playback songs. No differences in responsiveness were found (King and West, 1983a; West and King, 1985). We also know that females housed with members of other songbird species do not respond significantly differently from females cowbirds housed together (King and West, 1977). In view of these converging lines of evidence, we continue to have confidence that social housing does not interfere with the interpretation of playback data. The analyses required to clarify the issue indicate, however, some of the additional obligations incurred by adherence to the principle of social housing.

The experience of assessing the possibility of social influence between females also helped to highlight the strikingly reflexive nature of the female cowbird's copulatory response. Typically, females assume copulatory postures (if they are going to respond at all) before the conclusion of the male's one-second song. This characteristic has helped us to analyze what features of songs trigger responses (Figure 12.1). For example, during a song playback, females typically adopt copulatory postures before the whistle portion has begun, thus ruling out the whistle's structure as a key element in the female's perception of song (West et al., 1979). We have also used the female's playback response to study the role of experience in the male cowbird's development of song and the female's response to it. We probed for evidence of postnatal modifiability by housing acoustically naive North Carolina (NC) females with male cowbirds captured in Texas (TX) (King and West, 1983b). These males represent another cowbird subspecies, *M.a. obscurus*, whose range extends throughout the southwest U.S.A. We had already determined that the songs of the two subspecies differed and that females of both subspecies were sensitive to the acoustic differences we had previously identified. Each subspecies responded significantly more often to native song variants (King et al., 1980).

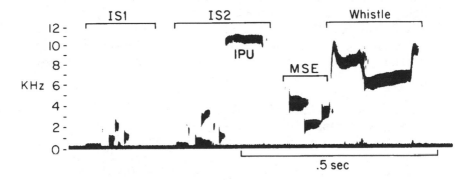

Figure 12.1.   An example of male cowbird song produced by a Texas *M. a. obscurus* male. The zero-crossings display represents frequency on the ordinate and time on the abscissa. The elements of the song are introductory syllables 1 and 2 (IS1 and 2), interphrase unit (IPU), mid-song element (MSE), and the terminal whistle. Songs of eastern *M. a. ater* do not contain MSEs, making this element a geographic marker between the two subspecies.

To test for female modifiability, we housed NC females with TX males for nine months, removed the males, waited another month, and then tested the females' copulatory responsiveness to Texas (TX) *M. a. obscurus* and (NC) *M. a. ater* songs. No evidence of modifiability was obtained. NC females housed with TX males responded as often to NC song as did normally reared NC females, and both responded at only negligible levels to TX song (King and West, 1983a). Thus, NC females appeared quite fixed with respect to song discrimination, apparently unaffected by the presence of female companions or by social associations with males. The close-minded nature of the female stood in stark contrast to the behavior of NC males. Our studies of male vocal development revealed them to be quite open-ended learners. Males learned songs either from tutor tapes or other males, and improvised on already learned songs in response to social and vocal stimulation (King and West, 1983 a, b, c; 1988, 1989; West and King, 1985, 1986). But males did not display equivalent amounts of modifiability in all social contexts; some circumstances were much more likely to motivate song modification than others. The key to predicting when alterations were likely to occur was knowledge of the social properties of the male's audience. We provide two examples.

Modifying Song Potency

Juvenile, as well as adults, show evidence of environmental sensitivity. To learn more about the motivating conditions under which adults change their vocal behavior, we carried out an experiment in which we controlled males' access to stimulation from males and females (West and King, 1980). We housed all the subjects in a common auditory environment. Five flight cages (2 x 2.5 x 3.5m) were arranged in a row. The ones at either end (and one identical in size across the room) contained one adult male and two females. These cages had black plastic covering relevant walls to prevent the males from seeing into other cages. Thus, the males were denied visual access to the activities of other subjects. The cages next to these each held one adult male and several females. The center cage contained three adult males and three females. Thus, the males in the middle three cages could see each other and the ones in the center cage could also physically interact with one another.

We recorded the songs of all the males and tested their playback effectiveness to a new set of females. Clear differences in potency occurred. The visually isolated males sang the most effective songs. Their songs elicited significantly more copulatory responses than the songs of the males in the center group or the males housed alone with females but allowed to witness the interactions of the group. The potencies of the latter two groups did not differ from one another.

What accounted for the differences? Why did the most "natural context," that of the mixed sex group, have a negative affect on potency? The most likely explanation was that the males living in physical proximity with one another learned to sing less potent songs because of the possibility of physical intimidation from other males. Observations in our resident colonies had confirmed that males cannot sing too potent songs unless they achieve a high rank in the male dominance hierarchy (West et al., 1981). Why did the physically isolated males, who only observed the group, also sing ineffective songs? We suspect that adult males did not need to experience an actual attack to learn about tune consequences of singing different songs. The males visually isolated from other males sang highly effective songs because they were spared the obligation of simultaneously courting females and competing with males. Thus, varying relevant social parameters in a simplified setting gave clear evidence of social sensitivity and suggested that males are stimulated to modify their behavior not only on the basis of auditory cues, but on the basis of social cues from their audience.

Modifying Song Content

In a series of experiments, we housed acoustically naive juvenile NC males with adult NC *M.a. ater* or TX *M.a. obscurus* females (King and West, 1983c, 1988). Because female cowbirds do not sing, we thus had another opportunity to look at male cowbird's responsivity to social, as opposed to vocal, stimulation—this time from a presumably less intimidating source. The results demonstrated that such housing led to multiple differences between the songs of the two groups of males. NC males with NC females sang prototypically NC song. But NC males with TX females included many song types in their repertoires that contained acoustic markers typically found only in the *M.a. obscurus* range.

Thus, the juvenile males demonstrated the capacity for nonimitative social learning. They did not learn to copy the behavior of their female companions, but to attend to the consequences. Because the effects on the songs were present long before the onset of courtship began, we knew that males were not exposed to copulatory responses from females. The differences between the groups emerged very early in development, long before females display copulatory postures or any clear courtship behavior. As early as November, before any definite or stereotyped song variants were

present, males housed with NC *M.a. ater* females sang significantly different vocal material than did NC males with Texas *M.a.
obscurus* females (King and West, 1988). Studying the song development of another group of naive NC males that were housed with
other species (canaries [*Serinus canaria*] or starlings [*Sturnus vulgaris*]) put the aforementioned findings in a clearer light. The
vocalizations of males housed with nonconspecifics contained a
little bit of everything, some prototypical NC song, some TX song,
and many other cowbird-like sounds. Their songs sounded typical
of cowbirds as a species but not of any particular population of
cowbirds. The males with other species developed generic repertoires tuned to no specific locale. Thus for males with female companions, it seemed nonimitative social learning was possible, a
process during which males learned by trial and error to sing the
songs that females attend to most often. But what were the
females doing?

   We set out to answer this question using several methods and
relying on considerable background knowledge about female
responsiveness to song overtures at other times of year, especially
during the breeding season. By virtue of our access to females in
different social environments, in this case, females residing in
indoor-outdoor aviaries in which they court and bred, we could
look at how females respond to songs from different males and different songs from the same male (West et al., 1981, 1983). We
observed that males and females interacted in diverse ways after
song overtures. Given the number of songs males sing (hundreds
each morning), "success" in the form of copulation was very rare,
perhaps every other day for a very active male. Sometimes females
only looked at males when they sang; sometimes, they lunged at
them; sometimes they flew away; and sometimes, they used "rat-
tle" vocalization. But a female's statistically most probable
response to a song by a male was to do nothing—she ignored him,
if we can be permitted an anthropomorphic interpretation.

   To learn about the proximate dynamics of social feedback
between males and females, we retreated from the complexities of
an aviary and returned to studies of males and females in small
enclosures, a setting in which we could best videotape males and
females observing and responding to each other. Analyses of the
videotaped interactions of males and females during the winter
and spring revealed that females responded with wing movements
in a small proportion of males' songs. One of the more striking
visual behaviors used by females during the winter and spring was

a wing stroke—that is, a rapid flicking of one or both wings that occurred during a male's song, typically 400 msec after song onset (West and King, 1988b; Figure 12.2).

When we first observed wing strokes, we surmised their significance based on the immediate behavior of the males (Figure 12.3). When females wing-stroked (the ratio of wing strokes to songs/hour was, on the average, one wing stroke for every one hundred songs), males approached the females and repeated the songs that had elicited in wing stroke, sometimes three or four times (West and King, 1988a). Thus, males appeared to notice the female's departure from her usual mode of inattention. To find out more about the meaning of wing strokes, we employed the playback procedure with another group of females in the subsequent year. We played back songs recorded in March that had or had not elicited wing strokes to females in breeding condition. Would the songs that had elicited wing strokes elicit more copulatory postures? The answer was yes. Wing-stroke songs were significantly more potent than songs sung in the same singing bout but that had occurred before a wing stroke. Songs sung after the wing stroke, quite often exact repetitions of the wing-stroke song, were also significantly more potent than songs sung before the female's display. Finally, we learned that songs eliciting wing strokes were the ones males tended to sing most frequently in the ensuing breeding season, well after wing stroking had ceased to occur. The retention of these songs suggested that the message of the wing stroke was not lost on the male in either a short-term or a long-term sense.

A final piece of information concerns the behavior of the female companions. Females were chosen for videotaping that differed considerably among themselves in one dimension: some had responded much more often in playback experiments than others. Although all the females had tended to respond to the same song types, some females responded to those songs much more often than other females. On an absolute scale, some females were more likely to be positively stimulated by song. Once we had data on wing stroking, we compared rates of wing stroking to playback responsiveness across the group of females. With one exception (a female who performed only two wing strokes), the more frequent wing strokers had been more responsive to song playback. Moreover, the potency of the males' songs was negatively correlated with female responsiveness: the males with the most structurally diverse and most potent repertoires had been housed with the females displaying the fewest wing strokes and the fewest play-

Figure 12.2. The middle panel shows a NC *M. a. ater* female cowbird producing a wing stroke while a male is singing. The wing stroke occurred approximately 500 msec after the male's one-sec. song began, shown in the upper panel, and was over as the song ended, lower panel.

Figure 12.3. An example of a male's reaction following a wing stroke. The male approaches the female and looks at her body and wing, which is now stretching, following the completion of the wing stroke.

back responses (King and West, 1989). Because we have data on only eight females, it is too soon to be sure that reactivity to song during development is a predictor of song responsiveness during the breeding season. But the data attest to the potential gains to be had by working with animals in several settings and developing longitudinal profiles of their communicative styles.

We are now at work analyzing a new signal system—one that may be the natural precursor to the male's knowledgeable use of vocal signals. It may be that males learn which songs to retain in their repertoire by social shaping on the basis of social cues from both females and males. The generic repertoires of males housed with members of other species suggests that males possess an overabundance of vocal material. The winnowing process toward locally appropriate songs (which did not happen in males with nonconspecifics) is a response to the male's social environment. It is a system in many ways analogous to a human's first signal system, which is not formal language but a variable collection of sounds, gestures, facial expressions, and body movements used by infant and care giver alike as they learn the pragmatics of communication during the first year of life.

SUMMARY AND CONCLUSIONS

Because it was environmental design, as well as experimental, that led to finding new forms of learning and communication in cowbirds, the data are offered as strong testimony in favor of the arguments on the value of seminaturalistic habitats put forth in this volume. We want to emphasize that we could only have obtained these findings because we could study birds longitudinally in a variety of settings. We needed studies in aviaries to show that females used song to choose mates. We also needed them to determine the proximate relationship between song and copulation. We also needed different, more secluded habitats to look at cowbirds during the times of year when social signals are less conspicuous. We also needed to be able to take these steps with the same individuals so that we could see how individual differences at one time of year translated into possible variation at other times of year (King and West, 1989).

In the study of songbirds, it is only recently that scientists have been able to see singing for what it really is, a social event involving a performer and an audience. And such a view necessarily involves different environmental requirements. A similar change in view is taking place with respect to the nature and interpretation of maternal imprinting in ducklings due primarily to greater attention to the natural ecology of ducklings during the period of early learning (Lickliter and Gottlieb, 1985).

In closing, we must issue the caveat that many of the specific remarks we have made apply to only a small number of avian species. We could no more issue guidelines applicable to the entire Order of Aves than provide one recipe with which to feed them. But the governing principles of researcher participation, social habitats, nested environments, and consideration of the experimental options versus experimenter obligations should have broad applicability as they represent approaches already in use by many investigators studying many different species. If the study of development is to continue to advance, it is imperative that the issue of habitat quality in the laboratory become a preeminent concern. Environments must not obscure the experience of ontogeny. All too often, shutting the door of an animals' cage has shut our minds as well.

ACKNOWLEDGEMENTS

The work reported here was supported by grants from the National Science Foundation. We thank E. Gibbons and an anonymous reviewer for their critiques.

13

Experiences with the Study of Rodents in Seminatural Enclosures

The goal of scientific study in comparative psychology and ethology is to develop a comprehensive understanding of animal behavior. The objective is to develop a solid empirical base, principles of generality, and comprehensive theories and explanations regarding the determinants of behavior. The study of animal behavior thus is primarily a basic science, although applied concerns are becoming increasingly prominent in some subdisciplines.

Following Tinbergen (1963), the study of animal behavior begins with careful observation and detailed description of the behavior and species under study. Four classes of questions are considered: (1) those involving mechanism, which concern the immediate, short-term determinants of behavior, including both physiological and environmental factors, (2) questions of development, which relate to the longer term ontogeny of behavior, (3) evolutionary questions, which are focused on the historical record and processes of behavioral evolution, and (4) questions of adaptive significance, which relate to the function of the behavior in facilitating survival and successful reproduction. A comprehensive understanding of behavior entails consideration of all four questions.

The focus of investigations in animal behavior, whether conducted in the laboratory or in the field, is on the natural lives and behavior of animals. With field research the problems of generalizing to the lives of animals in nature are, of course, minimal. However, many species, such as nocturnal, cryptic, rodents, are difficult to observe in the field. Also, many controls and manipulations are difficult or impossible under field conditions. For these reasons, much research is conducted in laboratory settings. The problem of validity in relation to the natural lives of animals becomes important in such research, whether the focus is on mechanism, development, evolution, or adaptive significance (see Dewsbury, 1987).

As a partial solution to the problems of combining experimental control with validity, many students of behavior have turned to seminatural enclosures. Problems of observation and control remain manageable, although not as straightforward as with animals in traditional, laboratory cages; yet animals have a more complex environment in which to live and behave. In general, by using seminatural enclosures we can combine the best, rather than the worst, of both field and laboratory approaches.

The purpose of this paper is to summarize some research conducted with rodents in seminatural enclosures in order to illustrate both the rationale for such research and the kinds of results that can be obtained. Consideration will be focused on the design of such enclosures as well as the value of seminatural enclosures as powerful situational determinants of behavior. It will be emphasized that behavior in such environments can be fundamentally different from that in small cages. Effects of enclosure experience that persist after removal also will be discussed. The review is intended to be representative rather than exhaustive; research from my own laboratory will be emphasized.

ENCLOSURES AND THEIR DESIGN

There is no simple definition of a seminatural enclosure. The kinds of environments to be considered here can be differentiated from traditional, small, bare test cages with respect to 1) increased size, 2) increased structural complexity, 3) increased social complexity, and 4) increased resemblance to natural conditions. Clearly, there is a continuum from small test cages at one end to large, outdoor enclosures at the other. In addition, studies conducted in enclosures often are continued for longer periods of time than those in small cages, thus permitting more stable behavioral and social relationships to emerge. In this chapter, I shall consider some research conducted in testing environments that differed from traditional small cages along all these dimensions. For simplicity, all will be termed seminatural enclosures or environments.

Many kinds of enclosures have been used by investigators of rodent behavior. The environments vary along many physical dimensions, depending especially on whether they are indoor or outdoor facilities. Outdoor enclosures maximize face validity, in that testing conditions are most similar to those in nature, while still limiting population size and excluding predators (Boice and Adams, 1980). Generally there are fewer constraints on size than with indoor enclosures. Although indoor enclosures mimic field conditions somewhat less closely than outdoor enclosures, control is increased as the experimenter can set the photoperiod, more easily exclude extraneous stimuli, and avoid the vicissitudes of climate and weather. In addition, indoor enclosures can be built near the work space used by investigators in fulfilling other responsibilities, and thus often can be monitored more regularly than outdoor enclosures.

An important problem in designing enclosures for behavioral research is to balance the problems of complexity and observability. In most enclosures the effective placement of various objects (e.g., baffles, barriers, nest boxes, or branches) can increase spatial complexity, mimic nature more closely, and provide functionally challenging environments. However, the layout must be designed in such a way as to maximize observability of the animals. Some researchers use mirrors or a video camera placed above the enclosures (McClintock and Anisko, 1982), whereas others look up at their animals through a Plexiglas floor (McGuire and Novak, 1984). The entire housing unit may be built of glass (Thiessen and Maxwell, 1979). Both glass and Plexiglas may require frequent cleaning and can be scratched; thus their use does not automatically ensure clear visibility.

Different sampling schedules are used for different purposes. Typically, one or more observations are made per day according to a temporal schedule. Procedures for objectively sampling the behavior of different animals have been systematized by Altmann

(1974). Videotape permits a more thorough sampling and repeated examination of important sequences of behavior. However, because of limited resolution, the size of the animal relative to that of the enclosure must be rather large if individual animals are to be identified. Use of telemetry can permit remote monitoring.

Some enclosures have been designed so that animals must make an operant response to obtain food and water, as well as to run or to change illumination levels (Goldstein, 1981; Kavanau, 1967b). An extraordinarily precise record of the animal's shifting motivational patterns can be obtained with these devices.

SEMINATURAL ENCLOSURES AS SITUATIONAL DETERMINANTS OF BEHAVIOR

The behavior of animals in enclosures often is closer to that in nature than is that in small cages. To illustrate the advantages of enclosures, I shall first review some research from other laboratories and then deal with some of my own work on this topic.

Research from Other Laboratories

Burrow Building. Rodents of many species dwell in burrows in nature. These provide protection from predators and the elements, as well as a safe place for the rearing of young. Thus, burrowing is an important behavior in the natural lives of these species. Domesticated rodents have been generally kept in small cages under the care of humans for many generations. An interesting question concerns the extent to which such domesticated animals will build burrows when given the opportunity; clearly, they cannot do so in small cages. In indoor testing enclosures, domesticated rats (Boice, 1977; Price, 1977), gerbils (Thiessen and Maxwell, 1979), and house mice (Adams and Boice, 1981) all dig complex burrows. Perhaps the most remarkable results were those of Boice (1977), who kept domesticated rats in an outdoor enclosure through a Missouri winter. The animals prospered throughout, built burrows typical of wild rats, and displayed many other behavioral patterns typical of nondomesticated rats in nature. Thus, when the situation was complex enough to permit the display of species-characteristic behavioral patterns, they were expressed even in animals that had undergone many generations of domestication.

Energetics. There is much interest in the energetic cost of reproduction; this is generally estimated in small-cage situations.

Schierwater and Klingel (1986) compared the energetic cost of reproduction in Djungarian hamsters in laboratory and seminatural conditions and found no significant difference. This may not be the case for other conditions. Further, exercise may enhance the health of the animal, as, for example, with respect to the cardiovascular system.

Foraging. A very active research area in recent literature has concerned the dynamics of foraging behavior and the extent to which animals optimize their foraging pattern with respect to benefits (e.g., food), and costs (e.g., energy, time, risk of predation) (see Krebs and McCleery, 1984). Because of the logistical difficulties of studying foraging under natural conditions, many laboratory studies have been conducted. In many such studies, such parameters as the time required to search for food have been simulated in operant chambers by using various schedules of reinforcement (Collier, 1982). These studies have been important in revealing many determinants of choice and in building cross-disciplinary interactions among scientists. However, several authors have questioned the validity of these simulations, arguing that greater spatial complexity may be critical in the study of foraging patterns (Baum, 1983; Shettleworth, 1989). Seminatural enclosures can be used to return the dimension of space and spatial complexity to the situation and to help to bridge the gap between small-cage simulations and the field (Mellgren, 1984).

Copulatory Behavior. There is a long tradition of testing the copulatory behavior of rodents in relatively small test cages. This has been an efficient procedure for the study of the effects of many experimental and physiological manipulations (Diakow, 1974; Sachs and Barfield, 1976). However, it is legitimate to question how results from such tests generalize to behavior in the field. Studies in seminatural enclosures provide a partial test of this generalizability. In a series of studies McClintock (reviewed in McClintock, 1981, 1984) has found copulatory patterns to be altered in enclosures. In general, she reported that females played a more active role in the regulation of copulatory behavior, and that the frequency and timing of various events in the male copulatory sequence appeared to have been altered. Although there are grounds for questioning some of the conclusions in these comparisons, it is clear that the test situation exerts an important influence on copulatory behavior. Situations in which the female can regulate the patterning of behavior reveal hormonal effects not apparent in simpler environments (Emery, 1986). Further, the study of the function of ultrasonic communication in the regulation of copulatory patterns has been facilitated by observing rats in enclosures larger and more complex than those used in traditional research (Anisko et al., 1978; White and Barfield, 1987).

In similar work, Hurst (1986) observed female house mice to display orderly patterns of moving to and from particular mating sites during episodes of copulatory activity. The use of a seminatural enclosure revealed spatio-temporal patterns of behavior that are eliminated in small cages.

Bruce Effect. The Bruce effect is the phenomenon wherein the pregnancy of a recently mated female rodent is blocked as a result of exposure to a strange male (Bruce, 1959). The phenomenon has received much study in a variety of species (Clulow et al., 1982; Milligan, 1976). It has been suggested that the Bruce effect may be a laboratory artifact (Bronson, 1979). Heske and Nelson (1984), however, studied the Bruce effect in prairie voles kept in 1.25 x 3 m outdoor enclosures, and found the Bruce effect even when the female could avoid the strange male and her mate was present. This increases the confidence one can have that the effect may occur in natural populations.

Mating and Social Systems. In nature, rodents live in a considerable array of different mating and social systems, ranging from dispersed, solitary patterns to patterns with animals in diverse social groups (Getz, 1978; Ostfeld, 1985). It is sometimes difficult to determine the nature of the mating and social system of small rodents in the field. In addition, the systems of some species vary with conditions; it is difficult to study these conditions if parameters cannot be manipulated systematically. Widespread use of seminaturalistic enclosures has been stimulated by the promise of information suggestive of mating systems in nature, combined with the possibility of introducing controlled manipulation of critical parameters.

The possibility of a monogamous mating system in prairie voles was suggested by Thomas and Birney (1979) after a study of pairing patterns in indoor pens. Field data appear to be generally consistent with this suggestion (Getz and Hofmann, 1986). Ågren (1984) found that Mongolian gerbils tested in outdoor enclosures lived in stable male-female pairs or in family groups. She has since studied the species in the field and found that the phenomena found in the laboratory generalized well to natural settings (Ågren et al., 1989).

The dynamics of long-term dominance and territorial relationships can be studied in various types of enclosures and cannot be duplicated in small cages. In a 15-month study, Adams and Boice (1983) found a two-stage pattern, beginning with relatively unstable and nonfunctional hierarchies in younger rats, and more stable and intense interactions associated with copulation in older males. Wolff (1985) found that, while one male house mouse was initially dominant in an enclosure, the younger males eventually deposed him and set up territories in the colony space. Wolff also reported that females moved freely throughout the space. Poole and Morgan (1976) found the dynamics of dominance and territorial interactions between male house mice to vary with the amount of available space.

In a classic study, Crowcroft (1966) examined house mice in a 20 ft. sq. room in which a set of barriers was used to provide spatial complexity. He learned much about the subtleties of the feeding and activity pattern, social behavior, and dominance interactions of this species. For example, dominant males that would attack subordinates outside the nest could be found sleeping with them when in the nest. Some subordinate males shifted their diurnal activity period to a time when the dominant male slept and behaved as a dominant to newcomers. In more recent research, Hurst (1987) used newer techniques, such as multivariate discriminant function analysis; social classes could be differentiated quantitatively and the relationship between class and reproductive success could be determined.

Once stable breeding groups of rodents (e.g., Norway rats) are established, there may be considerable resistance to strange intruders. The dynamics of these interactions, in relation to the nature of the colony defenders, the behavioral patterns displayed, and the conditions that can facilitate survival of the intruder have been analyzed in important enclosure studies (Blanchard and Blanchard, 1990; Blanchard et al., 1984; Lore and Flannelly, 1977; Lore et al., 1984). In general, it was found that alpha females attacked male intruders if resident males were not present. If situations are designed so that intruders have a refuge, conditions are closer to those experienced in nature than if the intruder is given no opportunity to flee from the residents' attacks. In an example of long-term research, Blanchard et al. (1988) observed colonies of rats every 100 days until all members of the colony died. A full life-span approach to the development and shifting dynamics of ago-nistic interactions thus was effected.

Crowding. In contrast to the research described above, which was designed to mimic the normal social behavior of rodents, other researchers have been interested in studying the effects of crowded conditions. This has largely been stimulated by an interest in modeling the effects of crowding in humans (Galle et al., 1972). The most systematic work in this area has been on Norway rats by John Calhoun and his associates (Calhoun, 1962a,b; Hill, 1987) in a variety of indoor and outdoor enclosures. Calhoun has found that under crowded conditions there are significant abnormalities in nest building, transport of young, eating, sexual behavior, and aggressive behavior.

Research from Our Laboratory.

Relevant research from our laboratory concerns studies of paternal behavior, dominance interactions in deer mice, and the major histocompatibility locus in house mice.

Paternal Behavior. Different species of voles (genus *Microtus*) appear to live in very different social systems in nature. Prairie voles, *M. ochrogaster*, and pine voles, *M. pinetorum*, live in family groups that often appear monogamous. By contrast, montane voles, *M. montanus*, and meadow voles, *M. pennsylvanicus*, are less social, live in individual home ranges, and rarely form stable social bonds (reviewed by Dewsbury, 1988a). The males of species living in family groups would be expected to display paternal care, whereas males of asocial species would not be expected to do so. Hartung and Dewsbury (1979) studied parental behavior in six species of muroid rodents in small laboratory cages. They found statistically significant differences in the paternal behavior of different species that were consistent with the behavior displayed in the field. Surprisingly, however, males of even the solitary species displayed substantial levels of paternal behavior, including sitting on the nest, and licking, manipulating, and retrieving pups. The authors concluded that paternal behavior in small test cages was not representative of that in the field; males of even asocial species forced into close contact with their litters displayed appreciable levels of paternal behavior.

The patterns are quite different when these species are studied in 1.3 x 1.3m seminatural pens (McGuire and Novak, 1984, 1986). Under these conditions male and female montane and meadow voles occupied different nests, and paternal care was essentially nonexistent. The study of paternal care patterns similar to those presumed to occur in nature appears to require a seminatural enclosure. Small-cage tests, at least as conducted so far, reveal the capacity for paternal behavior in asocial species, but appear to exaggerate the conditions for the display of paternal care. The precise factors responsible for these differences have not been determined.

Dominance and Differential Reproduction in Deer Mice. Our primary use of seminatural enclosures has been in a long-term study of dominance, copulation, and differential reproduction in deer mice, *Peromyscus maniculatus bairdi* (see Dewsbury, 1990b). Animals of many species appear to form stable dominance relationships, in which the outcome of agonistic interactions is generally settled with a predictable winner (Dewsbury, 1982a). There is much interest in the adaptive benefits of dominance (Dewsbury, 1982a). One prominent view is that dominant males gain differential access to females, and, as a result, sire more offspring (Dewsbury, 1982a). The data, however, are remarkably inconsistent on this point (reviewed by Dewsbury, 1982a). The hypothesis is important, as it concerns the evolutionary pressures related to a prominent pattern of social interaction. Proper tests of the hypothesis require that one adequately measure dominance, accurately record copulatory behavior, and determine the relative number of offspring actually sired. Deletion of any of these steps can lead to erroneous conclusions. There are relatively few studies in which all steps have been taken.

Deer mice are excellent animals for tests of the relationships among dominance, mating, and reproductive success because they adapt well to laboratory conditions. They occupy individual home ranges in nature, and appear likely to form stable dominance relationships with their neighbors (although they are unlikely to form complex dominance hierarchies). Females in nature sometimes mate multiply; multiply sired litters have been shown to be conceived in the field (Birdsall and Nash, 1973). We began research by studying groups in which one female and two males were allowed to mate in test cages measuring 48 x 27 x 13 cm; tests lasted several hours (Dewsbury, 1979). Dominant males attained significantly more ejaculations than subordinates, but there was no measure of paternity.

Dewsbury (1981) reported three additional experiments. The first was similar to that of Dewsbury (1979) except that a genetic marker was added. Females were homozygous for the blonde genotype; one male was of the wild genotype and one blonde. As the blonde gene is recessive, paternity could be determined unambiguously. As in the previous study, dominant males copulated significantly more than subordinates. This differential copulatory activity was not translated into differential reproductive success, however; dominant males did not sire significantly more young than subordinate males.

In the second experiment of Dewsbury (1981) a new paradigm was introduced. A seminatural enclosure, measuring 2.4 x 1.2 x 0.6 m, was built of plywood and Plexiglas. The base and three sides were of plywood, the top was of hardware cloth, and the front was of Plexiglas. Rocks and branches were placed around the edges of the enclosure. Food was dispersed about the floor of the chamber, and water was available through two bottles mounted at the back of the enclosure. The residents were two male and two female deer mice. The females were of the blonde genotype; one male was blonde and one was of the wild type. The animals remained in the enclosure for five days unless any animal appeared in danger of serious injury, in which case the run was terminated. Behavior was summarized during three 10-min samples each day, with all samples during the dark phase of the diurnal cycle, and separated from each other by at least 1 hour. If copulatory behavior was observed, the sampling period was extended. This basic paradigm was used in a series of studies (Dewsbury, 1981, 1983, 1984, 1988b, 1990a).

Perhaps the most critical result of the research in the enclosure was that, under these conditions, the dominant males not only copulated more but also sired more offspring, thus supporting the hypothesis relating dominance, copulation, and differential reproduction. This relationship, which was not apparent in the small-cage, short-term situation, was expressed in the longer-term, enclosure situation.

Equally remarkable were other changes in behavior which become apparent when the enclosure tests are compared to the small-cage tests. Females approached dominant males significantly more often than subordinates in the enclosures; this was not the case in small cages. Although this result suggests active female choice, other interpretations are possible, as dominant males were more accessible than subordinates. The ratio of chases to attacks was greatly altered in the enclosures, with chasing becoming more frequent relative to attacks; in the larger and more complex enclosures fewer chases ended in overt attacks by the dominant male. A pattern of spatial displacements also became expressed more clearly in the environments. Often, the subordinate male would give ground on the approach of the dominant male; no chase or attack ensued. All these changes suggest interactions with greater social discrimination and less intensive agonistic encounters. It is likely that social interactions in these seminaturalistic enclosures more closely resemble those in nature than do interactions in short-term tests in small cages.

In the third experiment, Dewsbury (1981) extended observations to 12 days, with results that were substantially the same as in the five-day study.

The actual patterning of copulatory behavior (mounts, intromissions, and ejaculations) was very similar in the enclosures to that of short-term tests with deer mice tested in small cages. However, the test situation was not an ideal one for the analysis of copulatory behavior, because the sampling procedure did not permit observation of all copulations. Further, comparisons of behavior in small cages and these studies are compromised in that, in the four-mouse, enclosure tests, changes in behavior may be due to social, rather than environmental, complexity. Sawrey and Dewsbury (1981) studied the effect of enclosure space on copulatory behavior in deer mice further, this time using single heterosexual pairs. Although the latency for male and female to come into proximity was greater in the enclosure than in small cages, the patterning of copulatory behavior was nearly identical in the two situations once copulation was initiated. It is clear that whereas some behavioral patterns are greatly changed in the enclosure, others remain unchanged. Indeed, many copulations in the enclosure took place on a flat rock in a corner of the enclosure. The surface of this rock was about the same size of a test cage, and the many ejaculatory series that occurred on the rock closely resembled those that occurred in small cages.

In a comparative study, Dewsbury (1983) examined the behavioral patterns of eight species of muroid rodents using a testing procedure similar to those used by Dewsbury (1981). Old-field mice (*P. polionotus*), a close relative of deer mice, showed a pattern of agonistic chases similar to that of deer mice. However, these patterns were different in the remaining six species. In various species, relatively little copulation or aggression was observed. Thus, a paradigm that was appropriate for testing the hypothesis in deer mice would require alteration if it were to be used to test the hypothesis with other species.

Similar data for Syrian golden hamsters in enclosures were reported by Lisk et al. (1989). Dominant males were active earlier in the day than subordinates and generally got to sleep near the females. Dominant males were more effective at penetrating the females' initial resistance and copulating with them. There were species differences in the effectiveness with which dominant male dwarf hamsters could exclude subordinate males from mating with estrous females (Wynne-Edwards and Lisk, 1988).

An additional study was designed to test the generality of a number of phenomena seen in small cages to the enclosure situation (Dewsbury, 1984). The procedures were similar except that the first male was introduced alone on the first day of the study, and the second male was introduced two days later. Females were introduced for limited tests on the afternoons of days 5 and 6. The most dramatic change was a sharp decrease in the levels of aggression when the males were in the absence of females. When females were introduced, however, the rate of male-male chasing increased to levels approximating those in the quartet situation in the earlier studies. The hypothesis that prior residence would affect agonistic interactions was supported; the males introduced first displayed significantly more chases than the second males. Although dominant males attained significantly more copulations than subordinates, the effects on differential reproduction were less clear, with dominant males appearing to sire significantly more according to some, but not all, appropriate analyses.

Male capacity for producing and delivering ejaculates appears limited both generally (Dewsbury, 1982b) and in these tests. When tested in small cages, males permitted to copulate to a criterion of sexual satiety (until 30 min without an intromission or 60 min without an ejaculation), on one day display markedly reduced copulatory behavior the next day. This result is important in estimating the reproductive capacity of males, and the potential for male choice as a result of these limitations. As in small cages, copulatory behavior was markedly reduced in the enclosures when females were introduced for a second consecutive day. Further, the levels of aggressive interaction also were reduced on the second day.

Sperm competition occurs when a female mates with more than one male during a single estrous period; the phenomenon has been studied in deer mice in small cages (Dewsbury and Baum-gardner, 1981). Patterns of sperm competition in the enclosures closely resembled those in small cages. Thus, many of the phenomena demonstrated in small cages could, in fact, be demonstrated in enclosures. Seminaturalistic enclosures can act as a kind of "halfway house" for testing the generalizability of phenomena found in small cages to natural situations.

In a later study, Dewsbury (1988b) analyzed the effects of male kinship and familiarity on the level of agonistic interactions. The procedures used were similar to those in previous research, except that in this and subsequent research, all animals were of the wild type, and paternity was determined with starch-gel electrophoresis using the transferrin locus as a marker gene for paternity. Males housed together continuously in small cages prior to introduction to the enclosure showed reduced levels of aggressive behavior in the enclosure whether or not they were siblings. Siblings housed apart showed no reduction in aggressive behavior compared to unrelated males housed apart. Clearly, familiarity is a more important determinant than kinship per se in modulating levels of aggressive interaction.

If dominance is to engender a differential selective advantage, dominance, at some level, might be expected to be heritable. Although there have been studies of the genetic bases of dominance, researchers generally have treated dominance as a unidimensional trait. Rather, dominance is a relationship between individuals. In the traditional sense of the term, such relationships cannot be inherited, at least not via a single genetic route. In a philopatric species such as deer mice, however, it may be reasonable to expect that the sons of dominant males may interact with the sons of subordinate males. One would therefore predict that the sons of dominant males should be dominant to sons of subordinates.

A genetic selection study was completed in which the base population was 28 quartets of deer mice tested in the five-day paradigm described above (Dewsbury, 1990a). After completing the five-day tests, the dominant and subordinate males each were mated during one afternoon in a small cage to nonexperimental females which were sisters of each other and not closely related to the males. The offspring in the two litters thus conceived, maternal cousins, were permitted to grow up without ever encountering their fathers. One male from each litter was then selected, so as to minimize the weight difference between them, and studied in the enclosure in order to determine whether dominance in the sons was predictable from that in the fathers. In general it was. In the first selected generation 15 sons of dominant males and 6 sons of subordinate males were dominant. The procedure was repeated in successive generations. The splits were 9–3 in the second selected generation, 7–1 in the third generation, and 6–1 in the fourth generation. Clearly, social dominance in the sons is predictable from that in the fathers, even though fathers and sons never encountered each other.

The same enclosures have been used in research with voles of the genus *Microtus* (Shapiro and Dewsbury, 1990). One-male, two-female triads of prairie voles and montane voles were studied for a 10-day period. As in small cages, prairie voles, which generally are monogamous in the field, showed more affiliative behavior, as they spent more time in side-by-side contact than montane voles, which rarely are monogamous. Patterns of changes in vaginal cytology also appear correlated with contrasting mating systems.

The Major Histocompatibility Complex in House Mice. Genes in the major histocompatibility complex (MHC) are among the most polymorphic coding loci known for vertebrates. Contrary to earlier views, it appears that these polymorphisms cannot be explained solely as a function of differential susceptibility to disease (Potts and Wakeland, 1990). However, it may be that differences in disease susceptibility become apparent when mice are tested under the more stressful social conditions found in nature and in enclosures, but not in small cages. Tests in small cages revealed that rodents can discriminate and display mating preferences among individuals bearing different MHC haplotypes (see Beauchamp et al., 1985). This suggests that mate choice may be one factor in maintaining the polymorphisms for MHC. However, such tests were conducted with highly inbred mice under highly unnatural conditions. To test various hypotheses, populations of house mice were established in large outdoor enclosures in Florida. Heterozygous males had a higher probability of acquiring a territory than homozygous males, and had larger harem sizes once established (Potts et al., 1989). These data from enclosures suggest that overdominance and reproductive mechanisms are involved in the maintenance of MHC polymorphism. Potts et al. (1991) found that, whereas MHC genotype had no detectable effect on neonatal mortality in the enclosures, mating preferences were strong enough to account for most of the genetic diversity found in house mice. Similar methods were used by Franks and Lenington (1986) to reveal the consequence of variation at the T-locus of house mice for social behavior and reproductive success.

EXPERIENCE IN ENCLOSURES PRODUCES LASTING EFFECTS ON ANIMALS

The research discussed thus far has been focused on the enclosure as a situational determinant of behavior; the studies discussed concerned the immediate effects of the environmental situation on behavior. Experience in seminatural enclosures also alters animals in ways that are detectable even after the animals are removed from the enclosures.

Developmental Studies

Perhaps the best-known, lasting, developmental effects are those that were demonstrated in a large number of studies concerned with enriched early experience. For example, Rosenzweig et al. (1972) studied both the behavior and neuroanatomy of rats reared in isolation versus enriched environments. In their most extreme environments, rats were maintained either isolated in small cages or in groups in outdoor 30 x 30 ft. enclosures. A variety of neurochemical and neuroanatomical changes were seen in the brains of rats reared in the enriched environments. Older animals also appear to benefit from experience in enriched environments. In further work, Greenough (1975) found detailed changes in the growth of dendrites and in synaptic connections in the brains of rats reared in complex environments.

Other developmental data have been collected on Mongolian gerbils (Clark and Galef, 1977, 1980). Gerbils reared under standard laboratory conditions were compared to those provided with a source of shelter to which they could flee. Those maintained under standard conditions showed accelerated eye opening, more rapid growth, earlier sexual maturation, and reduced adrenal gland size, when compared with animals reared in sheltered cages. They responded to novel visual stimuli by fleeing and displaying foot thumping, rather than by approaching the stimulus. Clark and Galef suggested that some of the effects usually attributed to genetic changes during the process of domestication may actually result from developmental processes attributable to rearing conditions in the laboratory. Similar results were reported for rats by

Boice (1981). Rats raised in cages, rather than burrows, appeared less fearful and sought shelter less than burrow-reared rats.

Studies with Adults

One of the research interests in our laboratory concerns mate preference. The process of choosing a mate can be of important selective consequences in the determination of levels of individual fitness. Ultimately mate choice can have substantial impact on evolutionary processes both via increased access to resources, which may be accrued by some potential partners more effectively than by others, and via the genes passed from the partner to one's own offspring.

We began our work in this area by trying to validate our apparatus and procedures with independent variables we were confident would affect choice, such as preferences for estrous versus diestrous females, and for females versus males. The results were much more complex than anticipated, as indicated below. A series of studies was conducted on the preferences of deer mice for bedding soiled by different conspecifics (Dewsbury et al., 1986). Both male and female mice consistently preferred bedding soiled by conspecifics over clean bedding. Thus, the apparatus seemed appropriate for the detection of differences. However, there was no consistent evidence that males preferred the odors of estrous females to those of females in diestrus. Male deer mice preferred bedding soiled by females for 6–9 days to that soiled by males for a comparable period of time. However, whereas normal, cage-reared males did not show a preference for female-soiled bedding when it was soiled for just 6–7 hr, males that had served in studies in the seminatural enclosure, discussed above, displayed such a preference. The experiment was replicated with the same result. It appears that the experience of spending five days in a mixed-sex group in a seminatural enclosure alters the discriminative abilities or preferences of male deer mice in regard to odors from females versus males. This finding appears reasonable when watching the animals in the enclosures, as mice come to attend very clearly to the gender of conspecifics, and treat males and females quite differently.

Related results have been attained with prairie voles (Taylor and Dewsbury, 1988). Males lacking sexual experience displayed no preference for estrous versus diestrous odors. Further, no preference for odors was detected in males that received sexual experience during monogamous cohabitation nor in those housed with two females. However, males that had either interacted with tethered estrous and diestrous females in a test situation or that had been housed with both males and females in a seminatural enclosure preferred odors from estrous females to those of diestrous females. Thus, male prairie voles differed from male deer mice in that reliable estrus-diestrus preferences were demonstrated. They were similar, however, in that experience in the seminatural enclosure with other males and females engendered preferences not apparent in other groups of animals. The experience of living in a complex social group in an enclosure seems to heighten social discrimination, or preference relative to control animals.

CONCLUSIONS

A major emphasis in the study of animal behavior in ethology and comparative psychology involves the study of behavior in relation to the natural lives of animals. The aims are to understand the mechanisms, development, evolution, and adaptive significance of the behavior of animals in relation to their natural lives. However, many treatments, behavioral patterns, and species are best studied in the laboratory. Seminatural enclosures provide a means for attaining some degree of validity in relation to the natural habitat while retaining appreciable control of significant variables. In some cases, such as the study of crowding and territoriality, such enclosures provide the only meaningful environments in which the behavioral patterns and processes can be studied. For others, such as studies of energetics and of copulatory behavior, the enclosures provide a "halfway house," in which the validity of behavioral patterns and experimental effects studied in small cages can be tested in more complex spatial and social situations. Some behavioral patterns, such as copulatory behavior, are minimally altered in enclosures versus small cages, whereas others, such as agonistic interactions, are greatly altered. Seminatural enclosures thus provide the only controlled way in which to study some behavioral processes and provide an important way of partial validation of laboratory phenomena in more naturalistic situations.

They constitute an important tool in the study of the behavior of rodents in relation to their natural lives.

ACKNOWLEDGMENT

Preparation of this paper and original research was supported by grants BNS-8520318 and BNS-8904974 from the National Science Foundation.

STEPHEN E. GLICKMAN
GLORIA S. CALDWELL

14

Studying Natural Behavior in Artificial Environments: The Problem of "Salient Elements"

All of us involved in the study of natural behaviors in artificial environments face a set of common problems when arranging conditions of housing and observation. It is nearly always impractical, and generally impossible, to simulate natural habitats in their entirety in the laboratory. Fortunately, it is probably also unnecessary (Hediger, 1969). Many biologically significant behaviors appear in captivity without the full range of stimuli provided by the natural environment (Boice, 1981). The challenge, in housing our subjects and constructing our experiments, is to find a set of conditions (the "salient elements") that provide sufficient support for the emergence of biologically significant results.

Any rational attempt to specify these "salient elements" must begin with an appraisal of those features of the natural habitat that could potentially influence the outcome of our research. We do not know of any simple formula. However, at a minimum, each experimenter must examine the conditions of housing and experimentation with the natural habitat situation in mind. Arrangements of food, water, temperature, lighting, shelter, topography, and the demography of living groups, should be thoughtful decisions, based upon the best available naturalistic data—not decisions based upon laboratory traditions developed with a limited set of domesticated subjects. Some features of the natural habitat (e.g., the presence of predators, or disease-bearing parasites) rarely will be introduced to the captive setting. Others (e.g., daily fluctuations in temperature) are routinely ignored, unless there is an experimental reason for their inclusion. In addition, there may be occasions where an experimenter deliberately distorts natural environmental conditions to answer a particular question. However, the considered inclusion, or exclusion, of the various "salient elements" is the best strategy for the production of ecologically valid results and humane animal maintenance.

This chapter is concerned with two general topics. First, we examine the translation of field-derived behavioral questions to the captive situation, including the kinds of decisions that are made involving the inclusion, or exclusion, of "salient elements" of captive environments. Second, since there are no simple formulas for such translation, we call attention to behavioral indicators that suggest successful construction of a seminatural environment and those that serve as indicators of failure.

The chapters in this volume by Dewsbury and Wyers contain ample evidence that the outcome of research with rodents is dependent upon the conditions of maintenance and testing. Although rodents will be featured in our consideration of these topics, we have not attempted to survey the literature. Rather, we have drawn upon relevant personal research experience. For some years, a number of researchers on the Berkeley campus have been involved in coordinated field-laboratory studies of woodrat behavior. In the course of this work, we have had occasion to make (correct and incorrect) decisions about which "salient elements" to include in maintenance or test environments. Our journey is probably typical, regarding the kinds of issues that will arise for researchers attempting to translate field-derived questions into laboratory research. However, there was one case where the conditions of housing had exceptionally powerful effects on the outcome of social encounters, one that is particularly relevant to the concerns of the present volume.

In addition, we have drawn upon personal research experience with an "excessive" behavior, appearing in Mongolian gerbils in captivity, to illustrate (1) the existence of behavioral indicators of failure to provide the "salient elements" of the natural environment, and (2) the kinds of research that can be undertaken to identify the missing elements.

FROM NATURE TO CAPTIVITY: STUDIES OF THE DUSKY-FOOTED WOODRAT

The dusky-footed woodrat (*Neotoma fuscipes*) is found in dense chaperral, woodland, and riparian habitats from Oregon to Baja, California (Ingles, 1965). It is a rodent of moderate size. Adult weights vary from 200 to 500 grams, as a function of age, sex, and subspecies. The presence of woodrats is marked by the construction of large "houses." These dwellings are assembled primarily from branches and twigs that have either been salvaged by the rats or gnawed from a shrub or tree and incorporated as part of the structure. The houses are often 3–6 feet in height, are most typically conical in shape (although there is much variation), and contain a complex array of interior passages and chambers (Figure 14.1). At least one of the chambers is lined with finely shredded plant materials, and presumably used as a sleeping area. The remaining chambers are sites for storage of a varied assortment of leaves, nuts, and fruits. Woodrats feed on a broad range of plant materials. They are excellent climbers and often forage in trees, or use elevated routes to move through their habitat. Houses are commonly found in clusters, but each house is occupied, at any one time, by an individual woodrat, except for mothers nursing pups.

Most woodrat researchers (Linsdaile and Tevis, 1951; Wallen, 1982) believe that houses and house-sites are a limiting resource for the members of any population. Possession of a house is essential for insulation from heat, cold, and rain. A house also affords limited protection from predation (for example, by owls, coyotes, cats) and provides a source of food when foraging is limited by rain or the presence of a predator. Finally, possession of a house carries with it access to the mating system. In this case, the problems confronting male and female woodrats are somewhat different. Males presumably attempt to maximize their reproductive contributions to the population by maintenance of a breeding-season territory that includes females and excludes other males. The situation for the female is considerably more complex. Woodrat litters are very small by rodent standards (modal litter size = 2 rats), although a given female may give birth to two or three litters in one breeding season (Linsdaile and Tevis, 1951). Infant woodrats weigh a substantial 14 grams at birth, and grow rapidly during 3 to

Figure 14.1.   Woodrat house near Shandon, California. This house was approximately four feet in height. Placement against a tree provided support, some additional shelter and the possibility of foraging on branches without descending to the ground.

4 weeks of nursing. The problem confronting each female is to maximize the number of her offspring that obtain a house, and there are a number of options. She ultimately makes a set of "decisions," ranging from maintaining her residence and encouraging the dispersal of her offspring, to leaving the natal house to a given litter.

Studying Woodrat Social Behavior in Captivity

As the result of trapping, tagging, and retrapping the members of population of woodrats over a period of months, Kurt Wallen (who was then a graduate student at Berkeley) had come to suspect that, although they resided one to a house, the members of a woodrat community knew their neighbors as individuals. If this were true, one would have expected very different social interactions between neighbors than with animals encountering each other for the first time. However, given the difficulties of observing social interactions between rats in nature, Wallen (1977, 1982) decided to translate the problem to a captive environment. The goal was the observation of social behavior between pairs of woodrats, to determine whether pairs of neighbors exhibited different patterns of behavior than animals trapped at great distances from each other. The first step was trapping woodrats at Wallen's study site near Pt. Reyes, California, and transporting them to the Field Station for Behavioral Research in the hills above the Berkeley campus. The next task was designing a captive environment which would support the essential aspects of behavior that occurred in nature. It involved the identification of the "salient elements" of the woodrat's natural situation and their translation to captivity.

The Maintenance Environment. Rats were maintained for several weeks in individual plastic cages, stocked with wood shavings, alfalfa, and an assortment of plant materials that were used for construction of a nest within the cage. Food was provided in the form of mouse chow, supplemented with sunflower seeds and various fresh fruits. Water was available ad lib. The rats were maintained in a room with an open window, permitting a natural cycle of sunlight and darkness, and variation in temperatures that closely approximated those found in their habitat.

The Test Environment. Wallen had inferred, on the basis of trapping patterns in his study area, that woodrats often met in neutral, unoccupied houses. He accordingly designed a large (2 x 1 x 1 meter) rectangular box to serve as a neutral meeting area. He accommodated the arboreal habits of these animals by providing a small network of elevated walkways on which the woodrats could interact.

What Was Done and What Was Found. After a single 30-minute habituation session in the test box, rats were paired, during the active part of their daily cycle, and their social interactions were observed under dim red light. Pairings involving female neighbors revealed different patterns of behavior than those involving females from remote areas. The strangers engaged in extensive investigatory behavior, while the neighbors did not. When neighbors were paired, one tended to become inactive, while the other explored the environment. Females also discriminated between males that lived in adjacent houses and males from remote areas. Males appeared to be less discriminating in their behavior, and Wallen obtained no evidence of differential behavioral interactions between males that were neighbors and strangers.

Of particular relevance to the present paper, active exploration of the test chamber during paired encounters were highly correlated with certain aspects of trapping in the natural habitat. Woodrats who were very active during social encounters in captivity were likely to be (1) trapped repeatedly on the same night, and (2) trapped at more than one house, more frequently than animals who were inactive in captivity. This kind of result, in which there were direct relations between dependent variables in nature and captivity, serves to validate the conditions of housing and testing, as does the differential behavior of the females. However, the lack of neighbor-stranger discrimination exhibited by male woodrats could either be a reflection of a true biological trait, or a failure of the test situation to provide the support necessary for the emergence of neighbor recognition in males.

Back to the Field

After the conclusion of Wallen's study, a more extensive field study of woodrats was undertaken near Shandon, California (Glickman et al., 1982). The population was visited on a monthly basis for several years. Individual animals were tagged, weighed, and examined for reproductive status. We also noted the houses at which each animal was captured. As the result of this work, a new set of variables was identified. In keeping with Wallen's results, trapping data suggested that male-female associations were relatively common, while male-male associations were very rare. However, as our long-term data accumulated, we realized that seasonal variation was a potentially critical influence on woodrat sociality. During the hot, dry months of summer, adult female woodrats are reproductively inactive and the vagina is imperforate. Analogous evidence of reproductive inactivity occurred in males. Testes were often retained in the abdomen and were of very small size. In contrast, during January and February every adult female in the population had a perforate vaginal opening and every male displayed large, palpable testes in a scrotal sac. Infants begin to appear in the population in March, continuing through April and May. These observations paralleled those reported by Linsdaile and Tevis (1951) in a study of woodrats near Carmel Valley, California.

The existence of a well-defined breeding season, with associated changes in hormonal substrates, suggested that there would be significant variation in social behaviors as a function of the time of year in which social interactions were observed. In particular, it seemed likely that male woodrats, competing for access to mates during the breeding season, would show enhanced levels of intrasexual aggression, when compared with male-male encounters during the nonbreeding season. Given the marked seasonal shifts in testes size, and the well-established correlation between testicular androgens and aggression in rodents (Brain, 1983), we also anticipated that seasonality in aggression would be linked to underlying changes in the hormonal substrate.

A Return to the Laboratory

In order to examine these possibilities, we initiated a study of the seasonality of social behavior in woodrats, with an initial focus on males. Animals were trapped, at sites near Carmel Valley, during the summer and during the breeding season, which began early in the year. In keeping with our expectations, wounds, presumably resulting from intermale aggression, were more frequently observed in woodrats trapped during the breeding season. Animals were transported to Berkeley and maintained in 10-gallon aquaria. Small translucent plastic tubes were placed in the aquaria and used by the rats as a place of retreat. Artificially illuminated day-night cycles were adjusted to approximate the natural day lengths appropriate to the season, and temperature were adjusted to a constant 70°F.

We tested animals in the situation developed by Wallen (1982), although we introduced the animals to the apparatus in small tubes that they used as a place of retreat in their home enclosure. This was done with the hope that the presence of these familiar tubes, in which they spent the majority of their daylight hours, would partially simulate the element of movement, from a familiar residence to a neutral area, experienced by woodrats in nature. However, a tube is not a house. Male woodrats fought vigorously during the winter breeding season, and displayed reduced levels of aggression during the nonbreeding months of late summer. Moreover, in intact male woodrats, seasonal variations in behavior were correlated with similar fluctuations in plasma testosterone. However, contrary to expectation, castrated woodrats continued to display seasonal cycles of aggression. Moreover, when these castrated animals were paired with gonadally intact males, there were no effects of castration on social dominance (Caldwell et al., 1984).

Faced with this surprising result, we reexamined the conditions of housing and testing. Although woodrats do "meet" in neutral houses, in the field there is reason to suspect possession of a house is critical to long-term survival in these animals, as well as being an essential component of a rat's ability to enter the breeding system. Therefore, a second study was undertaken to examine house defense in intact and castrated woodrats.

Maintenance and Test Environment. Six intact and six castrated male woodrats were housed in individual rooms at the Field Station for Behavioral Research near the Berkeley campus. Each 8 x 8 ft. room had one exterior Plexiglas and wire wall, which exposed the animals to natural light-dark cycles. Cool external temperatures were slightly modulated by the shelter, but there was substantial day-night and seasonal variation. Each floor was covered with wood shavings and six cubic feet of branches and twigs, in addition to an 8-foot long branch leading to an elevated platform. Every woodrat constructed a house of sticks within its room. Because we supplied an elevated platform supporting a triangular frame, and initiated construction with a few branches and wood shavings, animals were encouraged to construct their houses in a position that permitted observation of the interior. The woodrats completed the major portion of each house by piling additional twigs, branches and wood shavings on the exterior of the frame and carrying twigs into the interior (Figure 14.2). Each rat also constructed a "sleeping burrow" of woven bark that provided additional thermal protection inside its house. The rear of the frame was transparent and mounted flush with a window. An observer sitting in a small, darkened alcove behind the window could therefore see the interior of the house as well as the remainder of the room (Figure 14.3). Animals were provided with food ad lib (rat chow, supplemented with fresh fruits and vegetables) and water.

What We Did and What We Found. After rats had resided in their houses for at least three weeks, we initiated a set of 30-

Figure 14.2. "Houses" utilized by woodrats in captivity. Experimenters provided the triangular wooden frame, woodrats added the small branches and twigs, and rearranged the wood shavings. Subsequent models utilized plastic frames and supports in the interest of animal care regulations. However, woodrats would not build when we provided clean, smooth wooden dowels. House construction only proceeded when the natural rough wood product was made available to the rats.

minute tests. Pairings between intact and castrated animals occurred both in the intact's residences and in the residence of the castrate. Order of testing was counterbalanced with regard to gonadal condition and residence. An observer in the darkened alcove viewed the interaction under red light, during the evening hours that constituted an active portion of a woodrat's daily cycle. Aggressive interactions, consisting primarily of chases, sometimes accompanied by bites, were entered via a keyboard, permitting independent measures of frequency and duration of aggressive encounters for the member of each dyad.

Figure 14.3.   Interior of a woodrat house observed from a rear viewing window. The animal has constructed a secure sleeping area by carrying twigs from the floor of the room, up a ramp, and arranging them in the interior of the frame provided by the experimenters.

This procedure, which we carried out only during the breeding season, had a very different result from pairing in a neutral test box. The results of these encounters are depicted in Figures 14.4 and 14.5.

Intact residents were the aggressors more frequently ($t = 2.61$, $df = 5$, $p<.05$) and persistently ($t = 2.98$, $df = 5$, $p<.05$) than castrate intruders. However, castrate residents did not differ significantly from intact intruders in terms of frequency of aggression ($t = .95$, $df = 4$, $p>.20$), or persistence of aggression ($t =.66$, $df= 4$, $p>.20$).[1]

In retrospect, these results probably are not surprising. If possession of a house is a prerequisite to entering the mating system,

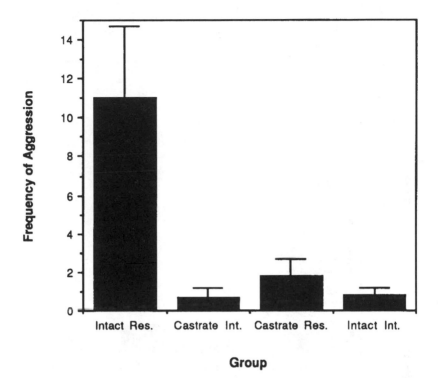

Figure 14.4. Frequency of aggression (Mean +/- S.E.) of woodrat Residents (Res.) and Intruders (Int.) as a function of gonadal condition (Intact or Castrate).

it seems reasonable that an intact male, capable of breeding, would be more likely to undertake the risks involved in aggressive house defense than a castrate male. In a subsequent (unpublished) study, we attempted to add another critical element of the natural situation to this confrontation: the presence of a mature female in the vicinity. Toward this end, individual females, in large wire enclosures, were housed for three weeks within rooms occupied by intact and castrated woodrats. One hour preceding the introduction of intruder males, we opened the doors of the females' cages and allowed male-female interactions to occur. Females were then returned to their cages, but remained within the rooms during the subsequent introduction of male intruders.

This proved to be a very poor simulation of the natural situation. Identifying a potential variable and introducing it in a biologically effective way are different problems. Although we have bred dusky-footed woodrats in captivity, in those cases each member of the pair had its own house and room, and a substantial period of

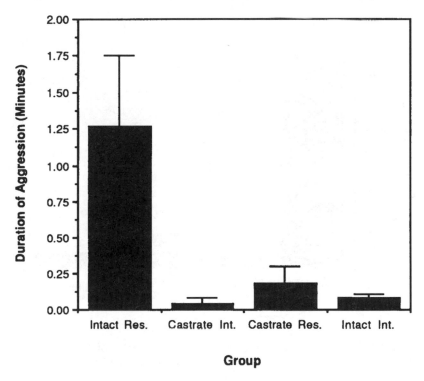

Figure 14.5.   Duration of aggression (Mean +/- S.E.) of woodrat Residents (Res.) and Intruders (Int.) as a function of gonadal condition (Intact or Castrate).

time to work out "relationships." Our attempts to short-circuit this spatially and temporally expensive procedure did not result in mating or affiliative interactions. Rather there were a series of investigatory and aggressive interactions between females and intact males, that were followed by highly variable interactions when intruder males were subsequently introduced into the residents' rooms. Our next step would be to return to the conditions that produced successful breeding—that is, housing males and females in separate rooms with passageways to permit visiting, as inferred from trapping patterns in nature. It would then be possible to control spontaneous interactions between intact males, castrate males, and females in a test environment that simulated the (presumably) "salient elements" of the natural situation in a more adequate manner.

Our woodrat studies provide a clear illustration of the powerful influence of the conditions of animal maintenance on the outcome of research. We began our work with a reasonable understanding of the natural history of woodrats, and selected problems and independent variables on the basis of field observation. However, the translation of a problem to captivity always involves a set of "guesses" about the minimum set of "salient elements" required to produce a biologically meaningful result. Our (and Wallen's) strategy initially involved construction of a relatively simple maintenance environment, taking into consideration the normal diet, daily fluctuations in illumination, and solitary sleeping habits of this species. The test environment, although more elaborate than the usual small, barren enclosure used for pairing rodents in a neutral arena, still lacked many of the elements found in the natural situation. The addition of "salient elements" to maintenance/test situations, in order to permit the emergence of new, biologically significant results, was carried out (sometimes successfully and sometimes not) by referring to the woodrat's lifestyle in nature.

INDICATORS OF THE ADEQUACY OF CAPTIVE ENVIRONMENTS

Although all maintenance environments must meet appropriate standards for humane housing, there is wide latitude in what constitutes an adequate environment for scientific study. Environments that contain the minimum elements for support of certain physiological studies may be totally inadequate for studies of behavior, and some behaviors may be more strongly buffered against distortions of the environment than are others. Although there are no simple rules for determining the adequacy of a maintenance habitat, there are at least four categories of behavioral indicators that provide general clues to the adequacy of maintenance.

1. The *appearance of fundamental, species-characteristic behavior patterns, following the same rules as those patterns in nature*, is a significant marker of appropriate maintenance. Breeding success, which is dependent upon functional mating and parental behavior, has been a widely used indicator of the adequa-

cy of a maintenance habitat. It is certainly a biologically significant reference point and has practical utility in programs designed
to protect the gene pools of endangered species. But other patterns
can also be employed. For example, we recently had occasion to
examine the meeting ceremonies, in captivity, of a group of spotted hyenas (*Crocuta crocta*), reared in peer groups. Participation in
meeting ceremonies, in which hyenas engage in mutual inspection
of the anogenital region, is a daily prosocial behavior of hyenas.
Kruuk (1972) had previously described the conditions governing
the emergence of meeting ceremonies. In particular he noted that
(1) meeting ceremonies were more probable under conditions of
social excitement; (2) meeting ceremonies were more likely to
occur between animals that had been separated for some period of
time; and (3) subordinate animals were likely to initiate meeting
ceremonies with dominant animals by offering their genitals for
inspection by the dominant. In a study involving controlled separation of individuals, and the excitement generated by release into
a previously restricted area of the enclosure, we were able to verify
all three of Kruuk's field observations (Krusko et al., 1988).

    Hyenas in our colony were also found to follow rules of scent
marking (i.e., deposition of paste from the anal scent glands) nearly
identical to those originally described by Kruuk (1972) for hyenas
in nature (Woodmansee et al., 1991). Such outcomes enhance one's
confidence in the adequacy of maintenance environments.

    2. *Significant quantitative/qualitative relationships between
behaviors in captivity and those in nature* can also serve as indicators of the adequacy of maintenance habitats or the validity of
assessment procedures. For example, we have previously noted the
correlation between measures of movement by woodrats in a laboratory test situation and patterns of trapping in the field (Wallen,
1982). This work involved direct measurement of the same animals in field and laboratory. However, it is also possible to gain
confidence in the adequacy of maintenance/test procedures by
comparing field and laboratory measures obtained on different populations, or with different species. For example, Carter et al. (1986)
presented convergent evidence from field and laboratory suggesting that prairie voles (*Microtus ochrogaster*) are generally monogamous, while Dewsbury and his collaborators have succeeded in
establishing behavioral correlates of monogamy (Dewsbury, 1981b)
or home-range size (Wilson et al., 1976) using interspecific variation. These various devices for relating field and laboratory indices

provide an alternative strategy for establishing the biological adequacy of maintenance/test environments.

3. *Behavioral stereotypes* have been described in zoological parks (Inhelder, 1962) and scientific laboratories (Sackett, 1968). They generally involve repetitive movements—that is, rapid pacing or rocking, but may extend to instances of self-mutilation, resulting from bites directed at the limbs. In general, such stereotypes are indicators of inadequate maintenance environments and are often linked with social isolation (of social species) or inappropriately limited spatial facilities.

4. *Excessive behaviors* (Falk, 1984) are yet another indicator of inadequate maintenance environments. These involve complete segments of an animal's consummatory repertoire (feeding, drinking, nest-building), which appear much more frequently than they would ever normally appear in nature. In the cases described by Falk, such effects (e.g., excessive drinking) are produced by the schedules of reinforcement. However there may be other, more subtle causes. For example, Gibbons and Stoskopf (1989) report that unusually high frequencies of "tail slapping" and of an upright vertical posture, permitting visual monitoring of the viewing public, characterized the behavior of bottle-nosed dolphins confronted by aquarium crowds of 30 or more visitors. Convergent evidence suggested that these behaviors were indicative of environmental stress.

SHREDDING OF NEST MATERIAL AND THE MONGOLIAN GERBIL

Some years ago, Stephen Glickman was involved in the study of consummatory behaviors in the Mongolian gerbil (*Meriones unguiculatus*). We typically maintained our gerbils in individual aquaria, on a substrate of gravel, and provided them with food and water ad lib, including fresh green leafy vegetables. We also provided abundant materials for the construction of nests, generally paper or cardboard, which was shredded into small strips. This is evidently a biologically significant aspect of gerbil existence. The construction of nests, within the sleeping chambers of their burrow systems, from shredded plant materials has been reported for gerbils in the natural habitat (Tanimoto, 1943). Bannikov (1954) has further noted that they will construct elaborate winter nests in Mongolia, supplementing the usual plant materials with wool, shed skin, hair, feathers, and cigarette papers (partially validating our use of paper nesting materials in laboratory studies).

We were intrigued with the ubiquity of shredding and nest construction. If one kept adding shredding material, gerbils would soon shred themselves out of living space. We determined that they preferred thicker, cardboard products to thinner papers (Glickman et al., 1967), that they preferred to visit portions of an enclosure that contained paper for shredding (Glickman, 1973), and that they preferred to shred index cards marked with the ventral gland scent of other gerbils than unmarked cards (Baran and Glickman, 1970). Ramon Blatt observed that gerbils would press a bar in order to receive strips of adding-machine tape, which were shredded and used for nest construction (cited in Glickman, 1973). However, there was a problem. If gerbils had shredded this much in nature, they would not have had time for all the activities necessary for a full gerbil life. It suggested that there was something inadequate about our maintenance conditions, and we set about determining the reason(s) for this excessive behavior.

First, we considered the possibility that the maintenance temperature was inadequate. Kinder (1927) had observed that rats maintained in an outdoor environment, at the American Museum of Natural History in New York City, varied their rates of nest construction as a function of external temperature. However, except for an initial drop in shredding when animals were first placed in very warm environments, there were no apparent effects of exposure to temperatures (ranging from 50° to 94° F) on shredding behavior (Baran and Glickman, 1970). Next we considered the possibility that the gerbils were merely acting to keep their incisors from growing in the face of a too-soft diet of Purina chow. However, when offered the opportunity to choose between a 3 x 5 in. index card for shredding, or a metal plate of the preferred thickness for gnawing, the gerbils clearly preferred to shred the index card.

We then considered the possibility that shredding, in nature, would be limited by competing activities. Depriving gerbils of food or water for 22 hours had no effect on shredding of nest materials. However, testing the animals in a situation in which they were compelled to choose between shredding and eating or drinking, reduced the time engaged in shredding behavior. This reduction only lasted for the time required to obtain a quick bite of food or drink of water. However, as Collier and his collaborators have emphasized, feeding, or ingesting water, in nature generally requires energetically expensive, time-consuming foraging, handling, and consumption (Collier and Rovee-Collier, 1983; Marwine and Collier, 1979).

Finally, we arranged a situation in which 48 adult male gerbils were housed under two different social conditions (individually or in groups of three) and in enclosures of two different sizes (12 x 12 in. or 72 x 44 in.). There were six subjects in each of the individual housing conditions, and six groups of subjects in each of the two enclosure-size conditions. After 4 habituation days, during which animals had access to index cards for nest construction, formal testing was initiated. Before each test, all paper (shredded and unshredded) was removed from each enclosure. Six 3 x 5 in. index cards, weighing 0.8 gms/card, were then placed in each enclosure. Twenty minutes later, all unshredded material was removed and six new cards were added. This procedure was repeated once again, until, in the course of one hour, each enclosure had received 18 index cards. This routine was repeated on 5 successive days. The amounts shredded by the grouped animals were divided by the number of animals (3), and each group was treated as a single sub-

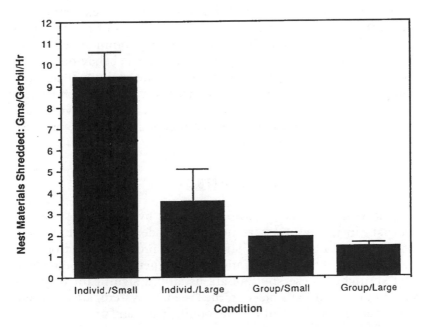

Figure 14.6. Shredding of nest materials in the Mongolian gerbil as a function of Individual (Individ.) or Group housing, and Small or Large living areas.

ject for statistical analysis. The results are presented in Figure 14.6.

Shredding of nest materials was markedly reduced in gerbils following provision of increased living space ($F_{1,20} = 11.12$, p<.005), or increased social contact ($F_{1,20} = 25.53$, p< .001). There was also a significant interaction ($F_{1,20} = 7.64$, p< .05) between the size and grouping variables; increasing the size of the enclosure did not result in a further reduction in shredding for the group-housed animals.

We know, both from earlier field reports and the recent studies of Ågren et al. (1989), that gerbils live in large social colonies. It makes biological sense that individually housed gerbils, deprived of the opportunity to forage, would find some way to fill their daily time-budget. In this case, shredding of nest materials played a prominent role. Roper and Polioudakis (1977) reached a similar conclusion, after observing little shredding of nest materials in a colony of gerbils observed under seminatural conditions in captivity.

More insightful animal behaviorists would probably have reached these conclusions without our elaborate experimental tools. But the method is there for the curious and the moral is there for all of us: if behaviors appear, which common sense dictates should not appear in such abundance, those behaviors can be used as indicators of missing elements in the maintenance environment. The decision regarding the necessity of adding those elements will be based upon the extent to which the excessive behaviors are likely to interfere with the problems under study, or cause unacceptable distress for the animals.

SOME CONCLUDING THOUGHTS

The vast majority of animals studied by biologists and psychologists in laboratories are nocturnal rodents. In nature, such animals commonly reside in shelters that protect them from climatic variation, or predation, and insulate them from the effects of direct sunlight. These shelters are usually a significant determinant of a given animal's place in the breeding system, are an essential element for successful rearing of offspring, and often provide a storage area for food. They are also constructed of material that are readily marked with personal scents and permit modification by the occupant. Yet few laboratory caging systems offer such rodents a place for personal retreat. The opportunity to retreat from light is rarely given as an option for nocturnal rodents in the laboratory, although it is always present for nocturnal rodents in nature. In the few experiments that have been done on this point, providing opportunity for retreat does not change the fundamental entrainment of hamsters (Pratt and Goldman, 1986; Rusak, 1975). The animals merely sample the illumination through controlled movements from a simulated burrow entrance. Provision of a darkened "burrow" area in which nocturnal rodents can retreat would seem to be a rational starting point for maintenance with such species, unless there is reason to force exposure to continuous light.

Clark and Galef (1977) have observed that Mongolian gerbils reared in a seminatural habitat permitting such retreat display hyperemotionality, compared with gerbils reared without such opportunities, and are more sensitive to movement in the environment. We have confirmed this observation in an environment providing a somewhat different burrow configuration (Spatz, Christenson, and Glickman, unpublished). These observations suggest that the relatively calm behavior of typical laboratory-reared gerbils is, in a sense, unnatural. It may be desirable in certain experimental contexts to have tame animals, but again the experimenter should be aware of the costs as well as the benefits.

Maintaining rodents on glass, plastic, or stainless steel substrates is sanitary and convenient. However, it facilitates cleanliness and disease prevention at the cost of eliminating normal shelter construction activities. Such housing may produce unnatural patterns of burrowing in corners of cages (Clark, personal communication), and ineffectual behavior indicating inappropriate housing. In addition, rodents deprived of their normal substrates may not scent-mark appropriately. It seems possible that too frequent cleaning may interfere with the recognition of self-produced marks that denote "home." Finally, many rodents use a dry, sandy substrate for bathing, preventing the accumulation of excessive oils produced by various exocrine glands (Eisenberg, 1963). Failure to provide appropriate substrates for such animals can have the effect of impairing health, as well as interfering with normal social behavior.

This is not meant to advocate abandoning all contemporary rodent housing for spatially extensive burrow systems prohibiting direct observation, and constructed of (literally) dirty substrates. That would be patently foolish. But it must be noted that, in opting for standard rodent housing, experimenters have chosen to ignore potentially "salient elements" of the natural habitat. There may be costs in such decisions. The primary goal of this paper is to encourage thoughtful review of maintenance/test environments with the natural habitat as a reference point.

ACKNOWLEDGMENTS

The authors are indebted to Edward Gibbons, Jr., and Irving Zucker for their helpful comments. Preparation of this paper was assisted by a grant from the National Institute of Mental Health (MH 39917).

NOTES

1. One castrated woodrat, after being wounded in a prior encounter, was unable to defend its residence against an intact intruder.

CHARLES T. SNOWDON

15

The Significance of Naturalistic Environments for Primate Behavioral Research

In the past few years there has been increasing concern about the conditions under which research animals are housed and the treatment they receive in research. Unfortunately, much of the public appears to believe that concern for the well-being of research animals is a new concern that has been forced on reluctant researchers by a group of persons with higher ethical values and finer moral concerns than those of the researchers. This is simply false. All competent researchers have always been concerned with the well-being of their research animals, and most researchers are motivated by extremely high ethical and moral concerns, choosing careers in research so that they can improve the welfare of human beings and other animals. One consequence of the new pressures for the control and regulation of research animal environments is that in the rush to write a single set of regulations critical species differences are ignored, and much of the history of the development of high quality captive research facilities is forgotten. Thus, we face the prospect of being regulated into captive environments for animals that in some instances may be less adequate than the environments they had been living in.

The focus of this chapter is the significance of naturalism (both physical and social) in captive environments for the study of primate behavior. I think that there are three lines of argument that support the value of naturalistic housing for research on primates whether in traditional research institutions such as universities or in zoological parks where increasingly there is an interest in quality research. One argument derives from the rules currently being formulated that require that those housing primates in captivity shall provide for their psychological well-being. I shall develop some possible criteria for evaluating psychological well-being and then argue that these criteria can best be met if primates are housed in seminatural environments in reasonable approximations to natural social groupings.

The second argument is derived from the goals of behavioral research. I shall argue that behavioral research on captive animals is most valid when paying close attention to the natural behavior of animals, and that only in naturalistic captive environments can the ultimate goals of behavioral research be met.

The third argument is more pragmatic. With the development of new techniques for studying behavior and physiology in naturalistic environments, one can obtain data more efficiently and can obtain data of higher quality while at the same time minimizing stress to animals.

Each of these arguments leads to the conclusion that naturalistic physical and social environments are to be preferred in behavioral research with captive primates.

NATURALISTIC ENVIRONMENTS AND PSYCHOLOGICAL WELL-BEING

The first problem is how to define and evaluate psychological well-being. This is a task that may keep psychologists and ethologists busy for several decades. *Webster's New Collegiate Dictionary* defines well-being in terms of happiness, comfort, and welfare, but these concepts are extremely difficult to evaluate in nonverbal organisms. We might use the absence of stress as a criterion, and measures of cortisol and more recently of immune system functioning (Coe and Scheffler, 1989) can be used to evaluate stress. However, for human beings some moderate levels of stress seem to be preferred to no stress at all. We seek to stimulate or test ourselves in our work and play, so psychological well-being may require that animals be allowed some moderate levels of self-controlled stress. Another criterion that could be used is the absence of abnormal behavior. However, unless we do field research on each of the species we study or study captive animals in naturalistic environments, we will not really know what behavior is normal and what abnormal, and what apparently abnormal behavior is a normal adaptation to captivity.

Another way to approach the definition of psychological well-being might be in terms of defining criteria for successful captive environments (Snowdon, 1989). I see three levels of criteria that can be applied to captive environments: (1) the veterinary criterion; (2) the biological criterion; and (3) the behavioral-ecological criterion.

The Veterinary Criterion

The veterinary criterion is the one most closely approximated by previous regulations of the National Institutes of Health and the United States Department of Agriculture. Animals must be provided with sanitary environments, with sterilizable cages, with clean, nutritionally adequate food and clean water, with a certain amount of air changes per hour, with regulated temperature, humidity and lighting cycles, and in minimal-sized cages. The goal of this criterion is to provide animals that are physically healthy in good nutritional status and free from disease. However, more recent versions of the regulations do recognize the need for social stimulation and acknowledge possible benefits of social housing (United States Public Health Service, 1985; USDA, 1991).

The Biological Criterion

The biological criterion is based on the Darwinian idea that reproductive success is the essential criterion of the viability of an individual. This criterion argues that in addition to the veterinary criterion given above, a good captive facility should also demonstrate high reproductive success. Thus, a captive colony must have many animals that are reproducing. For many of us doing primate research, this has been an implicit criterion during the last two decades. The natural habitats of primate populations of almost every species are severely threatened at present, and it would be irresponsible for a scientific institution or a zoological park to be dependent upon importation of wild-caught animals to maintain captive populations. Virtually every primate research facility in the United States has become self-sufficient in reproduction.

However, within the criterion of reproductive success there are two levels at which reproductive success can be evaluated. We can simply accept that success is achieved when a female delivers a live infant whether or not that female and her social group are able to rear that infant successfully. Unfortunately, some curators of mammals at zoological parks and many colony managers at research institutions are satisfied to hand-rear infant primates and feel that they have been successful. The argument is often used that hand-rearing of infants in rare or endangered species will lead to greater productivity, but this is rarely true. For gorillas, Beck and Power (1988) have shown that 71% of mother-reared captive gorillas are successful in rearing their own infants,while only 35% of hand-reared gorillas are successful in rearing their infants. Mother-reared captive born gorillas produce 2.5 times as many infants as do hand-reared mothers. Gorillas given social access in their first year of life show a higher conception rate than those deprived of social contact (63% versus 27%). Similar results have been documented for rhesus macaques (Arling and Harlow, 1967) and for Callitrichids (marmosets and tamarins; Epple, 1978; Snowdon et al., 1985; Tardif et al., 1984a,b). Furthermore, we know that for marmosets and tamarins there is an immediate postpartum ovulation and conception (Ziegler et al., 1987a). Hence removing infants does not speed up reproduction by the mother. The criterion of reproductive success becomes meaningful only if the animals themselves are able to rear their infants successfully. Hand-reared infants require a great deal of time and expense, and the resulting animals are apt to be deficient in parental care skills when they become reproductively mature.

Frequently, a more naturalistic environment will lead to better reproductive success in terms of parents caring for their own offspring. For the last decade we have been caring for a captive population of cotton-top tamarins (*Saguinus oedipus*), an endangered species endemic to Colombia. This species had been used in captivity for the last 25 years, but with relatively little long-term reproductive success. We began to analyze some of the factors involved in leading to reproductive success. It was known for related species of tamarin (the saddle-back tamarin, *Sagumus fuscicollis ssp.*, and the golden lion tamarin, *Leontopithecus rosalia*) that infants had to gain hands-on experience caring for someone else's infants if they were to become competent parents (Epple, 1978; Hoage, 1977). The same experience was necessary for cotton-top tamarins as well. Both in our own colony and in other colonies, the survival rate of infants born to parents who had had experience in caring for someone else's infants as they grew up was significantly greater than the survival rate of infants born to parents that had been hand-reared or who lacked early infant care experience in their development (Snowdon et al.,1985).

We have analyzed the care-taking patterns of different family members with infant cotton-top tamarins (Cleveland and Snowdon, 1984) and found a two-stage process of learning about infants. In this species the average interbirth interval is about seven months, with twins being more common than singletons. When new infants were born, the immediately older siblings (7 months of age) were not involved in carrying or caring for the infants. However, all age groups of siblings 14 months and older were equally involved in infant care. Male siblings and the father did a greater proportion of infant care than females in the first four weeks, with both sexes being equally involved after week 5. While the 7 month old infants did not carry or care for the new infants, they were the ones most frequently to initiate play with the younger siblings. We hypothesize that through play the 7 month olds learn how to interact with their younger siblings in an appropriate way, and that when the second set of new infant siblings is born, when the focal animals are 14 months of age, they then learn how to carry the infants. Since tamarins are arboreal primates, infants must be carried constantly until they are able to locomote independently. Tamarins that have not had early play and carrying interactions with young infants react to infants on their backs with a rejection reaction, attempting to push the infants off as soon as they climb on. This behavior is clearly not conducive to successful infant rearing.

The presence of older siblings who share in the carrying of infants significantly reduces the amount of time the mother carries the infants. Since the mother gives birth to twins that weigh 25% of her body weight at birth and has a postpartum estrus within three weeks of parturition that results in conception 85% of the time (Ziegler et al.,1987a), the mother has significant energetic burdens that are relieved by the presence of older siblings and a father to carry her infants. Thus, the presence of older siblings

increases at least the physical well-being of the mother and ultimately leads these siblings to be more competent parents.

What is the relevance of this example for the use of naturalistic environments? The minimum space requirement for a pair of
tamarins is a stainless steel cage about 0.5 m per side (United
States Public Health Service, 1985). This is not an environment in
which several successive sets of offspring can be left in order to
gain appropriate infant care experience. We surveyed all published
reports of colony success with cotton-top tamarins, and discovered
a correlation between cage size and complexity, and the reproductive success of the colony. Colonies that housed animals in small,
minimum standard cages had less than 50% of the reproductive
success rate of colonies that housed animals in larger, more complex cages (Snowdon et al., 1985). The most instructive data were
from two surveys of the breeding colony of the Marmoset Research
Facility at Oak Ridge, Tennessee. In 1978 they reported housing
animals in the small "standard" tamarin cages and had an infant
survival rate of about 25% (Gengozian et al., 1978). In 1984 they
reported on new caging procedures that had been instituted late in
the 1970s where breeding animals were housed in large family-
sized cages. Most other aspects of husbandry had remained similar,
but the reproductive success rate was 100% greater than it had
been previously (Tardif et al.,1984a,b). We think that one advantage of large complex cages is that it allows animals to be kept in
large family groups so that young animals can gain appropriate
infant-care experience. Other advantages to these naturalistic environments will be discussed later.

The Behavioral-Ecological Criterion

The third and most stringent criterion is the behavioral-ecological
criterion. This criterion is derived from the rationale used by those
of us who have developed captive breeding facilities for endangered
species. One of our goals, perhaps overidealistic, is that with captive breeding we are maintaining the species until such time as the
habitat can be rehabilitated and the animals reintroduced to the
wild. While there are several logical and practical problems with
the reintroduction of captive animals to a natural environment
(Brambell, 1977; Kleiman et al., 1986; Konstant and Mittermeier,
1982), the goal of reintroduction can serve as a means of evaluating
the quality of a captive environment.

The behavioral-ecological criterion of a captive environment argues that a good captive environment will be one that maintains in the captive population all of the motoric, social, cognitive, and other skills that would be needed by the animals to survive in the wild if they were to be reintroduced. There may never be enough habitat restoration to allow a reintroduction or there may be political and other barriers to release of animals, but we can ask the question of what skills animals would need if they were to be released, and then attempt to develop a captive environment that fosters these skills. Arguments based on the behavioral-ecological criterion have been made for nonendangered species by other authors (Coe, 1989; Bercovitch and Ziegler, 1989; Hediger, 1965; Maple and Finlay, 1989).

The consequences of adopting this behavioral-ecological criterion are many. First, the criterion requires that each species be treated separately. No general set of rules can be written to describe adequate captive environments for all species. Rather each species will have its own requirements for an adequate captive environment. Second, we will need to have a good knowledge of the normal behavior of a species in its natural environment, and we must have information on the ontogenetic mechanisms of behavior. If many of the important skills that an animal needs for survival are acquired through observational learning or social reinforcement, then we must provide such opportunities in captivity. Third, we must provide physical environments that will allow animals to express appropriate behavior. Arboreal animals will need to have structures where they can live and feed well above the floor of their captive environment. Finally, an animal living in a natural environment is exposed to unpredictable stressors such as a need to forage for food when it is scarce, or to detect and evade predators when present. And animals in the wild must be able to defend themselves against parasites and disease entities.

Thus, an ideal captive environment should provide animals with foraging skills; they should not just be given free food (Markowitz, 1979; McGrew et al., 1986). If predator avoidance must be learned, then captive animals might be given appropriate predator avoidance experience. Animals should not be kept in a totally sanitized environment, but should be exposed to low-level pathogens so that they can develop immune system resistance. Recent work indicating that housing animals in normal social environments results in improved immune system functioning (Coe et al., 1987; Coe and Scheffler, 1989).

Sensorimotor Skills

Providing an analogue to the normal physical environment can be important in the development of sensorimotor skills. For example, arboreal animals need to develop eye-limb coordination to move in their environment. To leap successfully from one branch to another, they must judge distances in three dimensions with great accuracy and evaluate how much "give" to expect when they land on a branch. Since direct pathways between two points are rare in the forest, arboreal animals must also learn appropriate detours. Clawed arboreal animals such as marmosets and tamarins will need to have a variety of "soft" structures such as branches, boards, and ropes that they can get their claws into while locomoting. Hard, impervious structures such as stainless steel are inappropriate and may even be dangerous. Thus, captive environments for marmosets and tamarins should have a complex of arboreal structures made of wood and fiber, with some structures fixed and unmoving and others with a great deal of flexibility. Different diameters of substrate materials will mimic the various diameters of structures in the natural environment.

Such naturalistic environments do work. Several years ago Dr. Anne Savage and I released four of our captive reared cotton-top tamarins to a natural forest at Monkey Jungle in Goulds, Florida, for a pilot study. Three of the four captive-reared monkeys showed excellent locomotor skills, being able to find their way through the canopy, to leap across gaps between branches in the trees, and to climb on branches and trunks of a wide range of diameters. We think our indoor caging environment comes close to providing the locomotor skills that tamarins need in the wild.

Social Learning

Even sophisticated biologists often assume that there are "instincts" for parental care or for predator recognition and defense, but with most primates, and with many other mammals as well, this assumption is unwarranted. I have already described the needs of tamarins to gain early infant-care experience with someone else's infants in order to gain parental competence themselves. There is no "innate" parental care system. Likewise, recent evidence shows that monkeys must learn to recognize predators as well. Mineka and her colleagues have published studies demonstrating that rhesus macaques (*Macaca mulatta*) must use observational learning to develop a fear of snakes. Mineka et al. (1980) showed that wild-caught macaques retained a fear of snakes even 15-20 years after being in captivity, while captive-born macaques showed no fear of snakes. Subsequently, Mineka et al. (1984) showed that captive-born macaques could learn snake fear by watching the reaction of a wild-caught macaque to snakes. The fear acquired through observational learning was quite robust, lasting at least six months, and the snake fear was resistant to several therapies for treating phobias (Mineka and Keir, 1983). More recent work shows that rhesus monkeys can even learn snake fear by watching videotapes of wild-caught monkeys behaving fearfully (Mineka and Cook, 1988). This result is quite exciting, since it might be possible to develop a library of video training types to preserve the behavior of the remaining wild monkeys to use to train captive monkeys.

We have tested cotton-top tamarins presenting them with live snakes, and found that fear or alarm responses in captive-born animals are no greater to a live boa constrictor than to a laboratory rat. Captive-reared monkeys showed a rapid habituation of response from the first to the second exposure to snakes, indicating a lack of innate fear of snakes (Hayes and Snowdon, 1990). In contrast, animals that could have seen snakes in the past displayed alarm and fear responses that did not habituate. Thus, if one wants to maintain predator avoidance skills in a captive population, it may be necessary to provide captive monkeys with opportunities to learn how to recognize predators and how to respond appropriately.

Another example of the importance of social reinforcement and observational learning is in the acquisition of vocal behavior. While the conclusion of a recent review of vocal development was that some primate vocal behavior appeared to be genetically determined (Newman and Symmes, 1982), recent studies indicate that both vocal structure, and appropriate usage may result from social learning. Seyfarth and Cheney (1986) found that young vervet monkeys (*Cercopithecus aethiops*) show a developmental course of both structure and usage that is best explained in terms of social learning from adults. If captive animals are to maintain the communication skills needed to function in the wild, they need exposure to appropriate social models during their development.

It is more conservative to assume that important aspects of behavior must be learned through social interactions than to assume that behavioral competence results from genetic and maturational variables. This means that naturalistic environments will be extremely important to provide young monkeys with the social and physical environment they will need to attain adult behavioral competence.

Foraging Skills and the Importance of Work

Good captive environments should also encourage the development of foraging skills. Prior to the reintroduction of the golden lion tamarin to its native habitat in Brazil, extensive training was done to prepare tamarins to find food that was distributed in space and time and that was frequently cryptic or hidden (Kleiman et al., 1986). Tamarins were trained to open birds' eggs, to handle whole pieces of food rather than cut-up food, and to accept food from the natural environment. McGrew et al. (1986) described a sap feeding device for use with common marmosets and pygmy marmosets, species that excavate holes in trees in the wild to obtain sap. This is a simple technique for maintaining sap extraction skills in captivity. Other studies have shown that foraging tasks can serve as an environmental enrichment tool (Chamove et al., 1982; Markowitz, 1979; Plimpton et al., 1981; Rosenblum and Smiley, 1984).

The enhancement of foraging skills leads to yet another important aspect of what animals need in a natural environment—good coping skills. How can we provide animals with coping skills so that they can respond to natural disruptions with adaptive behavior? A study by Mineka et al. (1986) showed that requiring foraging skills early in development yielded monkeys with greater competence and coping skills. Two groups of rhesus monkey infants were established in identical cages and were treated identically with one important exception. In one group the monkeys had to press levers, pull chains, or engage in other operant activities in order to obtain small food supplements or treats. The yoked environment had the same manipulanda, but acting on them brought no rewards. Animals in the yoked condition received free rewards in the same amount and at the same time as the monkeys who worked for their treats received theirs. This minor environmental manipulation had a profound effect on subsequent behavior. The monkeys that had to work for food showed less fearfulness in response to threatening stimuli and better coping responses when separated from other monkeys than did the monkeys who received free food. In addition there was a tendency for the "working" monkeys to have a lower cortisol response to these minor stresses. Thus, requiring animals to "work" seems to develop better coping skills.

At first blush it may seem paradoxical that work might lead to improved psychological well-being, but if we reflect on our own behavior, we obtain great satisfaction from work that leads to good results. Most of us would become quite maladjusted if we were deprived of useful work. Why not expect that monkeys need to work as well? Robert Yerkes (1925) in summing up his experience in rearing Great Apes in captivity mentioned the importance of work in developing a good captive environment. However, despite his advice from more than 60 years ago, many of us still think it is more humane to provide animals with all their needs without making them work. Yet in their natural environment animals must frequently work to find food, to guard against predators, to find shelter at night.

Why should work produce better coping skills? Mineka et al. (1986) discuss their experiment not in terms of work but in terms of control over the environment. They argue that animals with repeated experiences of having control over their environment (even so trivial seeming as pressing levers to obtain supplemental food) leads animals to develop confidence that they can control the environment in novel situations. It is the sense of control that provides the ameliorative effect of work. Hanson et al. (1976) also addressed the issue of perceived control in the well-being of primates. Adult rhesus monkeys were presented with loud, continuous white noise. Individuals in one group could press a lever to terminate the noise, while individuals in the yoked control group were exposed to the identical amount of noise, but had no means for terminating the noise. The group that was able to terminate noise voluntarily had significantly lower cortisol levels than those with no control. When control was subsequently removed from the first group, their cortisol levels increased to levels even higher than those of the animals with no control. In tests of social behavior, monkeys with perceived control over noise had low levels of aggression, those without control had intermediate levels of aggression, and the animals who had control but lost it had the highest aggression rates. These results taken together suggest that providing monkeys with environmental control can reduce stress levels, increase coping skills in novel situations, and reduce aggression. Thus, captive primates would benefit from opportunities to gain control over their environment, and these opportunities might be provided most readily in naturalistic environments where there is a greater variety of physical and social stimuli for the development of coping skills.

The adoption of the behavioral-ecological criterion for evaluating the well-being of captive primates means that animals must be housed in species-typical social groups in environments that provide analogues of important aspects of the natural environment for that species (Coe, 1989; Hediger, 1965). These naturalistic environments will enhance the sensorimotor skills of animals. They will increase the likelihood of social and observational learning of foraging skills, predator defense skills, communicative, reproductive, and infant-rearing skills with a minimal degree of effort from the human beings managing the colony. Finally, naturalistic captive environments can provide for psychological well-being by increasing the likelihood that animals will develop perceptions of control over their environment through work-like tasks. This experience should lead to less stressful, less aggressive animals who have better skills for coping with environmental change.

GOALS OF BEHAVIORAL RESEARCH

In addition to providing for the psychological well-being of primates, a naturalistic captive environment also serves to enhance the goals of behavioral research. One of the major journals of behavioral research, *Animal Behaviour*, requires as a constitutional charge to its editor that only "field and laboratory studies having a fundamental relationship to the natural lives of animals" be published. Thus, simply to publish in one of the major journals requires close attention to the natural behavior of animals.

I think there are three major goals of research on captive animals and each of these goals requires some degree of naturalistic housing for our animals. First, the data gathered on captive animals can be used to supplement and complement observations made in the field (Miller, 1977b). If data on captive animals are to be useful in completing or expanding upon field data, then these data must be gathered in relatively naturalistic settings. The second goal is to use the behavior of nonhuman primates to provide insights about the evolution or mechanism of human behavior. If parallels between animal and human behavior are to be valid, then we must have as a base the naturalistic behavior of both species. Finally, much of the value of behavioral research is comparative. How do different species respond to similar problems? If they respond in different ways, then it becomes of interest to seek the evolutionary, ecological, social, or ontogenetic differences between these species or populations in order to understand these differences. I also think that those interested in parallels between human and animal behavior should be just as interested in where parallels are not found. The diversity of behavioral responses displayed by other species to problems similar to those faced by human beings can illustrate novel ways of responses that could be used by humans that we might not have thought of otherwise.

There are many examples of the benefits of research with captive animals in complementing field studies. These benefits will be illustrated using three studies focusing on the cotton-top tamarin. Gautier and Gautier-Hion (1982) noted that much of the study of primate communication in the field is frustrated by the inability for humans to get close enough to monkeys to record intragroup or close-contact vocalizations. Thus, field studies on vocal communication are biased toward long-distance intergroup calls. In studies of vocal communication in captive cotton-top tamarins, we have been able to gain much more detailed descriptions of vocal structure and complexity of communication (Cleveland and Snowdon, 1982) than we have been able to obtain in the wild (A. Savage, unpublished data).

Even the long-distance intergroup calls might be more amenable to study in captive populations. The only field study of cotton-top tamarins described some aspects of intertroop encounters (Neyman, 1977), but there is much detail that is missing. McConnell and Snowdon (1986) developed a technique to simulate territorial encounters in captive groups. The technique was quite simple. Doors were opened between adjacent colony rooms, which had the effect of increasing the intensity of intratroop vocalizations from an unfamiliar group by 5-10 dB. After about 30 seconds of hearing these louder within group calls from an unfamiliar group, each group would alert to the direction of the other group, and begin a complex series of territorial displays that included piloerection, scent-marking, and a large number of vocalizations. Because we knew the age, sex, and individual identity of each animal, we could keep track of the roles played by different group members. Adults did all the territorial displays. Infants, juveniles, and subadults watched. There were certain calls and behaviors more often exhibited by females, and others more frequently exhibited by males. Thus, there was some division of labor by sex in the territorial encounters. Finally, we were able to describe sequences of behavioral escalation throughout the encounter and to document that different listeners reacted differently to different calls. Each group had different reactions to the behavior of its own group members versus the behavior of animals from the other group. It is unlikely that we would have been able to obtain these detailed descriptive data from observations of territorial encounters in the wild.

French and Snowdon (1981) examined how cotton-top tamarins responded to intruders. Male or female adult intruders were placed in small holding cages to protect them and they were introduced to pairs of adults. There was a sexual dimorphism of response to intruders. Females scent-marked to intruders of both sexes but showed little aggression. Males showed no aggression toward female intruders, but showed considerable aggression toward intruder males. We have subsequently found that some olfactory stimulus from the reproductive female in a group serves to inhibit ovulation in other females in the group (Savage et al., 1988), so the scent-marking response of the female might be appropriate. The results of the study suggest that in the wild there should be greater movement of females from group to group, and less movement of males. Some of the field observations of Neyman (1977) support this idea, as does a brilliant captive study of McGrew and McLuckie (1986). They attached flexible plastic ducting to home cages of cotton-top tamarins and ran the tubing through the halls to other colony rooms and to empty cages. Adolescent females were the first to explore the tubing, and that they traveled greater distances and were more likely to explore new environments than any other age-sex group in the family. Adolescent males were second, but the results fit closely with predictions from French and Snowdon (1981) and from field data (Neyman, 1977) that females should more likely be the dispersing sex. These examples illustrate how simulations of naturalistic environments can be used to learn more about the natural lives of animals than might be possible in nature.

The second and third goals of behavioral research both involve the value of comparative studies either for understanding human behavior or for understanding nonhuman species. We have studied the calls of pygmy marmosets (Pola and Snowdon, 1975) and found a series of trill-like calls that appeared to differ from each other on various parameters. Two types of trills differed only in duration but were used in very different contexts. Subsequently, we synthesized these trills and played synthesized versions of natural trills and intermediate forms of trills back to the pygmy marmosets and found evidence of categorical perception similar to how human subjects categorize human speech sounds (Snowdon and Pola, 1978). The category boundary was so precise that it appeared some neurological feature detectors must be involved. However, these results are strange, since they imply that monkeys (and humans) are not able to use all the information encoded in a vocalization. We replicated the categorical perception study with pygmy marmosets using syntheses that mimicked individual-specific features of the calls of monkeys in the colony. We found that animals paid close attention to these individual cues and made different decisions about category boundaries based on their previous experience with specific individuals (Snowdon, 1987). Thus, much of the information in a call is potentially usable if one tests animals appropriately. No one has yet tested human subjects to see if their categorical perception would be altered using socially familiar speech sounds.

French has repeated some of the studies he did with cotton-top tamarins using golden lion tamarins. In an intruder study that was almost identical in design to his cotton-top tamarin study, French and Inglett (1989) found that female golden lion tamarins reacted with severe aggression toward female intruders, while male lion tamarins were relatively calm about intruders of either sex. These results are exactly the opposite of those found with cotton-top tamarins. There are other interesting species differences. We have never found any signs of ovulation in subordinate cotton-top tamarins. Only the reproductive female ovulates (French et al., 1984; Ziegler et al., 1987b). However, in the golden lion tamarin both the reproductive female and subordinate females display synchrony in urinary estrogen peaks (French and Stribley, 1985). Thus, an intruding female lion tamarin might be a greater threat to the reproductive female than an intruding cotton-top tamarin female would be to the reproductive female. These results lead to some interesting predictions about species differences that should be studied in future field studies of both species.

In all these examples the use of natural social environments and the use of naturalistic techniques has produced interesting comparative results that can lead to interesting results and new predictions about human behavior and to suggestions for future field studies on related species of primates.

NONINVASIVE RESEARCH TECHNIQUES

One reason why naturalistic environments are often avoided for research is that to much of the research community, good research is synonymous with invasiveness. It is thought that animals must be kept isolated from one another under strict conditions of environmental control, especially for testing. While it is true that many research problems require that animals be tested under highly controlled conditions, it is possible to start thinking about alternative noninvasive research techniques. We have been using noninvasive techniques exclusively and we are able to ask interesting questions and obtain valuable data. More often than not, data can be gathered with greater efficiency than by using traditional methods, and we can obtain data of higher quality, because the methods of data collection no longer interact with the data itself. The third argument in support of housing animals in naturalistic environments is that higher quality data can be gathered with more efficiency in such an environment.

I will illustrate this argument with four examples of research with cotton-top tamarins. Because this is an endangered species, we have felt it especially important to avoid invasive data gathering techniques. However, these techniques can be used with a variety of species whether endangered or not.

One issue of concern to psychologists has been the evaluation of the learning ability or cognitive skills of monkeys. Several years ago Miles and Meyer (1956) used the Wisconsin General Testing Apparatus to study learning set acquisition in common marmosets (*Callithrix jacchus*). The marmosets showed no improvement in learning set-problem solving until after 900 learning set problems. Jeff Lande (unpublished observations) studied learning set formation in cotton-top tamarins in our colony, by bringing the apparatus to the animal's home cage. Individual animals readily learned to work at the apparatus and Lande found 90% performance in learning sets within 32 problems, comparable to the best data Harlow (1949) reported on rhesus monkeys, and far superior performance to that found in gorillas and chimpanzees. Menzel and Juno (1982) have reported rapid learning set performance in saddle-back tamarins.

Subsequently, we have used the apparatus to study color discrimination in cotton-top tamarins using Munsell color chips (Savage et al., 1987). We have been able to test animals in their home cages for 35-45 minutes a day, and in the color vision study individual subjects were tested for more than 3000 discrimination trials. Often in the middle of testing, animals will leave the apparatus to sit and groom with one of their companions for a few minutes and then return to complete testing. Testing in the home cage environment eliminates the need to separate animals from their social group and eliminates the long time of habituation to a novel testing environment that is traditionally used for testing animals. Thus, we have tested both learning capacity and visual perception by bringing an apparatus to the home cage.

We can also study auditory perception in the home cage environment. Bauers and Snowdon (1990) looked at the discrimination of two very similar calls in the repertoire of the cotton-top tamarin. Cleveland and Snowdon (1982) reported two forms of chirplike calls that differed only in the onset frequency and total frequency range. These chirps are less than 100 ms in duration, but they are used in very different circumstances. The F-chirp is a call used in territorial encounters between groups (McConnell and Snowdon, 1986) and the G-chirp is used in relaxed social situations when animals move about undisturbed. Do these calls provide sufficient information for animals to discriminate between these contexts or do the animals make use of other contextual features? We presented 8 pairs of tamarins with examples of these two chirp types through speakers placed in their home cages and found that they could discriminate between the chirps in the absence of any other contextual differences. The monkeys made responses that were appropriate to each call type. The behavioral responses that indicated a discrimination between the two calls were significantly different on the first test trial. This is in marked contrast with other studies of discrimination of monkey vocalizations using operant techniques (Petersen et al., 1984) where it appears that approximately 30,000 training trials were needed to establish a discrimination. Playback testing in the animals' natural social environment is considerably more efficient.

We became interested in evaluating the reproductive status and function of our tamarins on order to manage the colony effectively and in order to know more about their reproduction. Traditionally, endocrinologists have used serum measurements for hormonal studies, but this creates several problems. First, with small primates repeated blood draws rapidly deplete red blood cells and can put the animal at great risk. Thus serum measurements must be made infrequently. Second, the process of taking serum samples is often stressful and interferes with subsequent behavioral measurements. To solve this problem we have developed several methods for measuring steroids and gonadotropins in urine rather than serum (French et al., 1983; Ziegler et al., 1987a).

We have made use of the arboreal habits of our monkeys, and the fact that humans and other primates frequently urinate upon waking up each morning. We now walk into our environments each morning as the monkeys wake up and by feeding the individuals whose sample we want and holding a container underneath the animal, we can collect the first morning void without having to capture or even handle the animal. Without large naturalistic environments we would not have been able to collect urine samples noninvasively. Using this technique we have been able to obtain continuous hormonal measurements for 7-10 months at a time, including samples on the day of parturition and during lactation without compromising the female's health or that of her infants (Ziegler et al., 1987a,b). We can make behavioral observations immediately after urine collection, and we have been able to provide the most complete description of hormonal changes over a pregnancy cycle and during puberty of any primate species, including human beings. The development of a noninvasive technique for collecting and measuring urine has allowed us to gather better and more complete reproductive data than are available for other species. We have recently modified these techniques to extract hormonal data from fecal samples, which allows us to study hormonal changes in wild populations. We are currently collecting samples from wild cotton-top tamarins in Colombia, and our assay techniques will be useful in evaluating the reproductive status of the wild tamarins.

Each of these noninvasive techniques has allowed us to gather research information that is more detailed or more efficiently obtained than would have been possible with traditional techniques. However, much of the successful operation of our techniques is a direct result of the naturalistic environment in which the animals are housed.

SUMMARY

Three lines of argument support the use of naturalistic housing for primate behavioral research. The first argument was based on considerations of psychological well-being. The behavioral-ecological criterion for captive environments requires that these environments be as naturalistic as possible. The second line of argument was based on the goals of behavioral research, which use captive animals to illustrate natural behavior that might otherwise be difficult to analyze in nature, and to develop comparisons between primates and humans and between species of nonhuman primates. These goals require for the most part that the behavior we study be as natural as possible, expressed in a natural social and environmental context. Finally, there are several potential noninvasive research techniques that can be used in naturalistic environments that can lead to high quality research data without stressing the animals. There is significant value in the use of naturalistic captive environments for behavioral research on primates.

ACKNOWLEDGMENTS

The research described here and the preparation of the manuscript were supported by a USPHS research scientist award MH 00,177, and USPHS research grants MH 29,775 and MH 35,215. Funds from the World Wildlife Fund and the National Science Foundation BNS 89–22741 are supporting field work in Colombia. I am grateful for the colleagueship of Kim Bauers, Jayne Cleveland, Margaret Elowson, Jeffrey French, Sheryl Hayes, Patricia McConnell, Anne Savage, Tina Widowski, and Toni Ziegler in developing our approach to naturalistic housing and the research done with our marmosets and tamarins. I thank an anonymous reviewer for a helpful critique of a previous version of this chapter.

MELINDA A. NOVAK
PEGGY O'NEILL
SUE A. BECKLEY
STEPHEN J. SUOMI

16

Naturalistic Environments for Captive Primates

*A cross-section of nature is not an equivalent
part of the whole, but merely a piece which,
on being completely isolated, alters its quality.
In other words: Nature means more than the
sum of an infinite number of containers of space
[cages], however natural. . . . Naturalness in
the treatment of wild animals does not consist,
therefore, of a pedantic imitation of one model
section of nature . . . but of an adequate transposition of natural conditions.*
Hediger, 1964

Within the last decade, researchers have come to appreciate the importance of laboratory housing environments in altering, reducing, or exacerbating the behavioral and physiological responses of nonhuman primate subjects. This new awareness coincides with the growing influence of socioecology and field research in characterizing the complexities of primate behavior and with an emerging concern about primate conservation. As a result, some scientists are developing laboratory habitats that are "more natural" and that presumably improve the quality of life of captive animals and promote their well-being. In addition, new federal regulations now raise the issue of "psychological well-being" as a major factor in defining legally suitable laboratory environments. However, the connection between "naturalistic settings" and "well-being" of captive nonhuman primates has not yet been explicitly evaluated .

One of the goals of naturalizing indoor laboratory environments is to restore the broader behavioral repertoire of wild animals who are now maintained in captivity and to expand the behavioral repertoire of laboratory-born animals who have never experienced a "real-life" environment. To what extent do laboratory environments designed to be more natural actually promote primate well-being? In particular, can laboratory-reared, socially naive monkeys with limited exposure to environmental challenges benefit from a naturalistic environment?

The purpose of this chapter is to consider the concept of naturalistic laboratory environments for primates. We begin by examining the broad range of cage environments that might be considered naturalistic. Next, we will evaluate current laboratory simulations of selected features of natural environments, with a special emphasis on cage size, cage complexity, foraging devices, tool-using implements, and social housing. Finally, we will consider the relationship between naturalistic settings and primate well-being.

NATURALIZING PRIMATE LABORATORY ENVIRONMENTS BY HOMOLOGY OR ANALOGY

Laboratory environments for captive primates are often viewed as stark and harsh, having little resemblance to life under more natural conditions. The primates in these environments are typically healthy in the physical sense, neither malnourished nor parasite-ridden like many of their wild-living counterparts. On the other hand, they are seldom exposed to the daily challenges of outdoor life (e.g., foraging, escaping predation), and their behavioral repertoire may be quite limited. A primary purpose of creating a naturalistic environment is to reproduce a part of nature and the challenges that it carries. However, it is no simple matter to make laboratory environments naturalistic, and an exact replica of the natural environment can probably never be produced indoors or even outdoors in corrals or fenced enclosures (Hediger, 1964). Given this caveat, we will define a naturalistic primate environment as a transposition, either by duplication or by modeling, of some features of the habitats of wild-living primates. A laboratory environment that literally duplicates certain features of the wild represents a transposition by *homology*. In this case, the primary focus is on the physical characteristics of the environment (e.g., providing foodstuffs only found in the wild, adding trees). Regardless of the outcome, this type of transposition looks natural and is usually pleasing to the public, if not to the primates and scientists as well (Coe, 1989).

On the other hand, we might use the term "naturalistic" to describe laboratory environments that model the functional consequences of some of the features of the wild environment without directly duplicating them. Here the focus is on the behavior of laboratory primates rather than on the replication of physical features per se, and the task is to provide these animals with some of the challenges experienced by their wild-living counterparts, albeit in somewhat different form. Such environments represent transpositions of nature by *analogy* (e.g., artificial devices used to promote foraging behavior or jungle gym structures that emulate tree branches). This kind of transposition does not necessarily look natural and may not be appreciated fully by the public. Attempts to make analogous simulations more natural (i.e., creating life-like trees from artificial materials to replace jungle gyms) may be important in zoological gardens to provide the public with a sense of life in nature but may seem less necessary in laboratory settings as long as the requisite species-normative patterns of behavior are elicited and maintained.

Potential pitfalls exist with both methods of naturalizing captive environments. In using homologous transpositions, one cannot assume that natural materials will always elicit a broader range of species-normative patterns of behavior. Naturalistic laboratory environments should be more than mere window dressing; to be effective they need to stimulate primates behaviorally, physiologically, and cognitively. In analogous transpositions, the lack of similarity in form may interfere with the expression of the desired functional outcome. Researchers need to establish that general principles present in the wild environment apply in the modeled case. One excellent example is provided by field/laboratory comparisons of foraging behavior in the blue jay (Kamil et al., 1988; for primate examples, see Andrews and Rosenblum, 1988; Menzel and Juno, 1985).

Although we have presented these hypothesized transpositions separately, a naturalistic laboratory environment could be a hybrid of both homologous and analogous simulations of selected environmental features. Homologous transposition may be more practical for small arboreal primates, where tree branches, vine tangles, and food items can be used to create an indoor habitat that closely resembles a part of the natural environment (e.g., prosimian primate housing at the Duke Primate Center; see chapter 9 by Izard and Pereira, this volume). However, given the rate at which many larger primates, such as macaques, destroy natural materials, an analogous transposition may be more practical with these species.

SIMULATIONS OF THE ENVIRONMENTS OF WILD-LIVING PRIMATES

Whether homologous, analogous, or some combination, successful simulation of the wild environment in laboratory settings requires at least three steps: (1) identifying the natural environment of the species in question, (2) selecting the features of this environment that can be modeled either by homology or analogy in the laboratory, and (3) proving that the outcome of such modeling is an increase in species-normative patterns of behavior.

What constitutes the natural environment for a particular primate species? Is it the environment in which the species evolved, as has been frequently suggested? If so, this notion poses some problems for those who create naturalistic habitats. Given the rate of habitat destruction in various parts of the world (particularly tropical forests), the environments to which individuals of a particular species were adapted may no longer exist. As these environments disappear, species that are unable to adjust to new conditions face the threat of extinction. For example, intensive logging of tropical forests has reduced the range of suitable habitat for the golden lion tamarin by 98% (Mittermeier et al., 1986; Kleiman, 1981), and strong conservation efforts will be required to prevent its ultimate demise (Kleiman et al., 1986). In some cases, then, the environment in which a species evolved is no longer present and homologous simulations may be more difficult to engineer in the laboratory.

On the other hand, there may be a wide range of natural environments in which a particular species is found. Generalist species such as rhesus monkeys are successfully adapted to a wide variety of habitats, raising the question about which environment should be modeled in the laboratory. Reasonable arguments can be advanced for selecting the most common habitat, the habitat that produces the highest reproductive rate, the habitat that is least encroached upon by humans, the habitat occupied by the rhesus monkey's most recent ancestral species, or any habitat in which the species is found. Rhesus monkeys are particularly interesting because, in contrast to golden lion tamarins, they seem to thrive under disturbed conditions. They prefer to feed on the plants that appear following deforestation, and they also flourish in areas of human habitation (Goldstein, 1981). These characteristics led Richard et al. (1989) to call the rhesus macaque a "weed species" because it both competes with and depends upon humans for food resources. Thus, there is usually greater flexibility and more options in simulating the environment of a generalist or "weed" species such as the rhesus monkey, than a specialist such as the golden lion tamarin (i.e., a variety of analogous transpositions may be available to the researcher).

Once the natural environment is identified, certain features can be selected for simulation. Such simulations will necessarily be limited by the characteristics of the laboratory environment (e.g., amount of space present) and by various requirements of federal and institutional regulatory agencies regarding husbandry and treatment of laboratory animals (e.g., the use of imporous structures made of steel rather than wood, even though wood is more naturalistic both by homology and analogy). In this paper, we will focus on five general aspects of laboratory settings that have been studied with regard to their potential role in simulating natural environments. These are: (1) cage space and its influence on locomotor activity and aggressiveness, (2) nonmovable elements such as cage partitions and perches as they relate to climbing and segregation of animals, (3) food distribution as an aspect of foraging, (4) provision of manipulable objects as it relates to tool using, and (5) group composition as it pertains to social dynamics.

Space

There is no question that primates housed indoors or even in large outdoor enclosures are restricted spatially compared to most wild-living animals. Although wild primates do not range with complete freedom throughout their habitat (i.e., limitations may be set by terrain, location of food, water and sleeping sites, social relationships, or the presence of other species), their home ranges usually encompass a wide area. Calculations of the distance primates typically cover in a given day (i.e., day range) vary from a low of 30 m for the slow loris to a high of 13 km for the hamadryas baboon (Clutton-Brock and Harvey, 1977). Obviously, these ranges cannot be reproduced in a homologous manner in the laboratory. Thus, an analogous laboratory simulation at a minimum should provide adequate opportunities for locomotion and other whole-body movement (e.g., swinging, brachiating, walking, running) in both vertical and horizontal planes. At issue is how much space is necessary to elicit these movements—and is absolute amount of space the most relevant variable?

Few researchers have systematically examined the relationship between activity levels (as measured in terms of locomotor and whole-body movements) and cage space. Increases in cage space have been associated with increases in locomotion (as noted in lorises by Dashbach et al., 1977) and with a normalization of locomotor patterns (as in rhesus monkeys; Paulk et al., 1977). Although sifakas showed heightened locomotor activity when moved from smaller indoor to larger outdoor enclosures, this effect may have resulted more from outdoor living or novelty than space per se (Macedonia, 1987). However, increasing cage space does not always result in increased activity. A juvenile gorilla moved with his social group from a smaller outdoor enclosure to a larger, more natural-looking, outdoor environment showed reductions in activity and increases in self-directed behavior that persisted across a four-year period (Goerke et al., 1977). Furthermore, increasing the cage size of individually housed rhesus monkeys had virtually no effect on heart rate or activity (Line et al., 1990). Inadequate cage space, on the other hand, seems to be partially responsible for the development of abnormal motor patterns such as back-flips, rocking, and pacing (Berkson, 1967; Draper and Bernstein, 1963; Goosen, 1981).

Although cage space may not always correlate with activity, it is assumed to play an important role in modifying tensions between socially-housed conspecifics (i.e., cage space should be inversely correlated with levels of aggression). This suggests that primates in captivity are more aggressive than primates in natural settings, an assumption not necessarily confirmed by existing data. For example, primates in natural settings are known to be particularly aggressive to one another, especially during the breeding season (Lindburg, 1971) even though they have access to a large spatial area. Furthermore, in direct comparison tests, laboratory monkeys were not more aggressive than their semifree ranging counterparts (Novak et al., 1992).

In the laboratory, the relationship between cage size and aggression is unclear. Changes in the spatial dimensions of captive primate housing do not always support the notion that "more is better." A reduction in cage space has been associated with both an increase in aggression (Alexander and Roth, 1971; Elton and Anderson, 1977; Nash and Chilton, 1986; Southwick, 1967) and a decrease in aggression (Erwin, 1979). In two studies, an increase in space actually led to an increase in aggression (Erwin, 1979; Novak and Drewsen, 1989). Thus, attempts to increase cage space and make the cage environment more natural may not always yield positive outcomes.

Structural Nonmovable Components

A cage is more than empty space filled only by animals. It can include internal structures such as partitions and furnishings. Researchers are now realizing the potential importance of these cage components in simulating features of the natural environment and in promoting species-normative patterns of behavior (Eisenberg and Kleiman, 1977). Many cages are now designed to reflect the interrelationships between the morphology of an animal, its behavioral propensities, and its natural habitat (Stuart, 1981).

Partitions or Barriers. Hediger (1964) was the first to introduce the notion of functional space—the space that animals use rather than the total amount of space. In general, primates more readily exploit cage space, which is configured by barriers and partitions as compared to open space (Hancocks, 1980; Hutchins et al., 1978). In the wild, natural barriers such as bushes and tree trunks provide good hiding places and enable primates to escape from the aggression of other animals, to breed surreptitiously (Drickamer, 1974), and to spend some time either alone or with a subset of other animals.

Solid partitions in laboratory cages can function in similar ways. Visual restriction in laboratory settings can reduce aggressive behavior between animals (Chamove, 1984). Monkeys housed in indoor/outdoor pens, for example, often escape from an aggressor by moving through the swinging doors to the other side of the pen. However, barriers or partitions can occasionally produce undesirable outcomes. Harem groups of pigtail macaques were more aggressive to one another when given access to two rooms connected by a small door than when kept in a single room without solid partitions (Erwin, 1979). Erwin suggests that the reduced ability of the dominant male to resolve disputes in the two-room situation as compared to the single room may have accounted for the effect. Thus, under some conditions, solid room dividers may actually be associated with higher levels of aggression.

Solid partitions may also play a critical role in breeding efforts, especially if copulations between certain partners require some degree of separation from other animals. Despite the existence of dominance hierarchies in free-ranging troops of rhesus monkeys, subordinate males can successfully copulate with females if they are out of the view of more dominant males (Drickamer, 1974). At the University of Massachusetts Primate Lab, cage dividers enabled a low ranking male to breed with females when and only when he was on the other side of a solid partition from the dominant male (Novak, unpublished observations).

Opaque movable partitions may be used to simulate dense canopy environments and create "new" areas in which animals aggregate and coexist. The cage environments of two arboreal species, the common marmoset and the cotton-top tamarin, were altered by the addition of hanging screens made of cotton material. After these changes, both tamarins and marmosets showed increased affiliation (McKenzie et al., 1986).

Furnishings. Furnishings are nonportable structures that also increase the functional space available to captive primates. Such furnishings include nest boxes, perches, suspended branches or poles, swings, and ropes and can be made of either natural or artificial materials. Perches and suspended poles can be used to emulate tree branches and may be particularly important for species that spend a majority of their time in the trees.

The actual arrangement of these structures can be modified to meet the needs of particular arboreal primates (Stuart, 1981). Dusky leaf monkeys, *Presbytis obscurus*, are relatively sedentary tree dwellers that move quadrupedally in a horizontal direction through the canopy. Banded leaf monkeys (*P. melalophos*), in contrast, are more active and forage vertically in the trees by leaping from branch to branch. As noted by Stuart (1981), broad poles that are mainly horizontal in orientation could be used to simulate the natural microhabitat of the dusky leaf monkey, whereas banded leaf monkeys would need a more complex system of smaller poles and branches located in both a horizontal and vertical direction. The success of such pole arrangements can be determined by space utilization studies. In one such study, the tarsier, a vertical clinger and leaper, preferred midlevel heights and small diameter poles located in the vertical plane to other poles of varying sizes and orientations (Roberts and Cunningham, 1986).

Perches can also be important for terrestrial species such as macaques, functioning as sleeping areas off the ground and escape routes during aggression. When multiple perches are provided at different heights, they enable monkeys to segregate into smaller social units. Thus, nonmovable structural cage furnishings provide researchers with the opportunity to construct a wide variety of laboratory living units and to simulate features of many different habitats of both arboreal and terrestrial species (O'Neill, 1988). Space utilization measures can be used to determine whether such simulations actually elicit the relevant behavior patterns and preferences.

Foraging Activities

Wild primates spend a significant amount of their waking activity foraging for food. They frequently compete with other animals for access to food and subsist on a wide variety of food types, which may require different search and preparation strategies. In captivity, primates are typically fed once or twice a day from a constant food source that involves little collection or preparation. This contrast between field and most laboratory environments makes food presentation an excellent area for naturalistic simulation.

Simulations of foraging activity have taken two forms. In the more homologous transposition, the act of foraging for a particular natural food substance is virtually re-created. For example, wild marmosets feed on exudates by gouging wells in gum trees and licking the sap as it oozes out. When laboratory-reared marmosets were exposed to wooden dowels containing sap reservoirs, they displayed the correct posture, dug appropriately shaped wells, retrieved the gum substance, and then urinated in the wells, exactly as observed in the wild (McGrew et al., 1986).

In the other approach, researchers have attempted to simulate the more general components of foraging behavior such as the search for and the preparation of food. Such strategies include making food more difficult to find, altering the distribution of food in the cage environment and/or requiring specific movements from the animals using food as a reward. For example, searching behavior can be elicited by placing food in some kind of substrate (e.g., deep woodchip bedding). The outcome of this manipulation is generally beneficial in that it can increase activity and decrease aggression (Chamove et al., 1982), increase exploration and decrease social interaction (Byrne and Suomi, 1991), or make laboratory animals more closely resemble their free-ranging counterparts (Novak et al., 1992). However, some of these beneficial effects may be lost if the foraging demand is too high or the food restricted to only a part of the environment (Plimpton et al., 1981).

The density or distribution of food may have important ramifications for foraging simulations involving socially housed primates. Under laboratory conditions, food is usually provided in highly clumped patches (e.g., the food is contained in one or two small food hoppers or it is placed on the small floor surface). Troop-living primates seldom encounter this kind of distribution for all their food sources in nature. The continuous provision of clumped food is often undesirable in captive primates, because it can lead to increased agonistic behavior and decreased social contact (Boccia et al., 1988; de Waal, 1984; Southwick, 1967; Wrangham, 1974). It may also promote differential food consumption and energy expenditure in high and low ranking animals.

A potential solution to this problem, one that involves more naturalistic foraging, is to alter the delivery and/or distribution patterns of food in the cage environment. Bloomsmith et al. (1988) observed a decrease in aggression, competition, and abnormal behavior in chimpanzees that were provided with food items requiring lengthy processing or search time. However, not all alterations lead to decreased competition. The practice of providing group-housed primates with chopped rather than whole fruits and vegetables, for example, did not reduce competition. Instead dietary diversity, time spent feeding, and total amount consumed was actually higher under the whole food condition (Smith et al., 1989).

Not enough attention has been focused on the effects of distributing the food more widely throughout the cage environment of group-housed primates. In addition to the horizontal spread of resources along the cage floor, food can also be distributed vertically along cage walls. Beckley and Novak (1989) employed this strategy by creating foraging racks that could be attached to the mesh sidewalls of group cages (Fig. 16.1). Rhesus monkeys could then collect food at different heights and locations in a simulation of more natural feeding behavior.

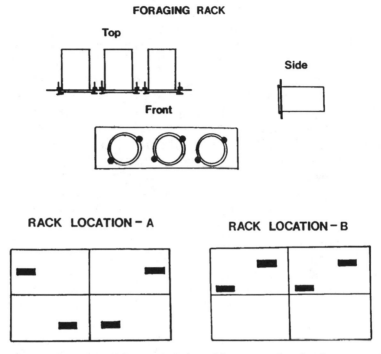

Figure 16.1. Diagram of foraging rack and location of racks during condition A (racks located at all levels and heights) and condition B (racks located in upper half of floor-to-ceiling cage).

Extensive observations clearly demonstrated the positive benefits of foraging from racks. In contrast to the standard feeding condition (i.e., food scattered on the floor along the edge of the cage), monkeys were more active, hoarded food less often, and therefore expended more energy retrieving food from the racks, especially when racks were positioned high off the ground (see Fig. 16.2). A major cause of tension during feeding—namely, competition between group members—was also changed under this condition. The racks served to spread animals out, and no single animal could successfully monopolize more than one rack. As an added benefit, monkeys showed generally higher levels of affiliation in the period immediately following rack but not standard feeding.

Thus, relatively minor changes in food presentation can have major effects on activity and tension in captive groups of primates.

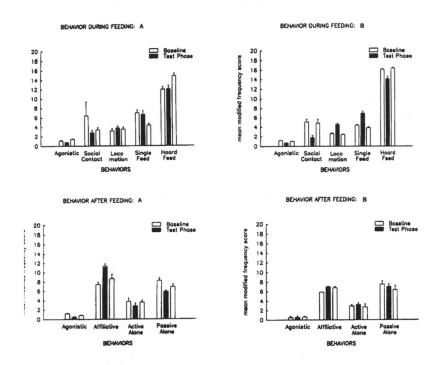

Figure 16.2. The effect of foraging racks on behavior as determined by a comparison of the test phase (rack feeding) with baseline (standard feeding with food scattered on the floor). During condition A, racks were located at all heights on the cage front. During condition B, racks were placed only in the upper half of the cage.

Manipulable Objects

Wild primates are generally curious animals who explore and manipulate a broad range of stimuli in the natural environment. However, these stimuli are generally not available in laboratory environments. Instead, researchers provide animals with a variety of manipulable objects (buckets, blocks of wood, rubber rings, hoses, chains, or balls; cf., O'Neill, 1988) and creatively designed enrichment devices (food puzzle boxes, seed feeders, grooming boards; cf., Bayne et al., 1991) to foster curiosity and exploration in laboratory housed primates. Manipulation of such objects may lead to food reward, as in the case of grooming boards, or the manipulation may be an end in itself. Most objects intrinsically elicit some kind of response from nonhuman primates. For example, introduction of manipulable objects increased activity in captive orangutans (Tripp, 1985) and increased manipulation in capuchin monkeys (Visalberghi, 1988; Westergaard and Fragaszy, 1985), lion-tailed macaques (Westergaard and Lindquist, 1987), squirrel monkeys (Eterovic and Ferchmin, 1985; Fragaszy and Mason, 1978), and chimpanzees (Bloomstrand et al., 1986).

Interest in objects, however, can wane with familiarity (Line, 1987; Miller and Quaitt, 1985; Paquette and Prescott, 1988). The rate of this decline will vary, in part, with the complexity of the stimulus, its composition, and with species propensities and individual preferences (Eterovic and Ferchmin, 1985). To maintain curiosity, primates may require new or "recycled" objects on a regular basis, or stimuli that serve to elicit and maintain species typical patterns of behavior (e.g., grooming boards).

The maintenance of curiosity in lab animals, however, may depend upon the quality of the environment in which the animal is housed. In contrast to socially housed monkeys, individually housed monkeys may be less motivated to manipulate objects and may habituate to such objects more rapidly (Line et al., 1991; Novak, unpublished observations). However, objects or devices that specifically elicit species-typical activity (e.g., branches or grooming boards) can be effective in eliciting responses in many individually housed monkeys (Bayne et al., 1991; Reinhardt et al., 1987).

Primates also possess special cognitive abilities, which may remain underdeveloped in standard laboratory housing conditions. Wild chimpanzees, for example, craft tools and use them to harvest new or difficult to obtain food sources such as termites. Wild capuchins are opportunistic feeders and employ active search strategies designed to exploit hidden food sources. They strip bark and crack open nuts (Izawa, 1979; Izawa and Mizuno, 1977; Janson, 1985). Researchers are now examining manipulable objects as potential devices not only for fostering curiosity and exploration but also for promoting tool-using behavior and problem solving activities in different species of primates.

Although tool using and tool making may develop in laboratory situations where objects are merely provided, more frequently primates may require specific challenges or opportunities to elicit these patterns of behavior. Termite fishing behavior has been modeled for laboratory chimpanzees (Keeling et al., 1988). Captive capuchins will employ tools in captivity to solve complex tasks related to obtaining food. Capuchins have used wooden blocks to crack the hard shell of nuts, manufactured probes from browse for obtaining liquid through small openings, and employed paper towels as sponges (Visalberghi, 1986; Visalberghi and Vitale, 1990; Westergaard and Fragaszy, 1987). However, not all species are avid tool users. Squirrel monkeys, tested in similar situations, never showed any evidence of tool use (Westergaard and Fragaszy, 1987). Whether laboratory primates benefit from exposure to tool-using tasks may depend on species-characteristic foraging patterns. Tool using may be more prevalent and more readily elicted in species that exploit a wide variety of food sources (Visalberghi and Antinucci, 1986).

Social Stimulation

The vast majority of primate species are social rather than solitary, and an enormous variety of social organizations has been recorded across the Primate Order. Some animals live as monogamous pairs (e.g., marmosets and gibbons), whereas others are members of large multimale troops (e.g., macaques and baboons). Even presumably solitary species such as orangutans make occasional social contacts, and female adult orangutans are seldom ever alone because of the prolonged developmental dependence of their young. Thus, housing individuals in species-typical groupings may be an important part of naturalizing laboratory environments for virtually all primate species (Novak and Suomi, 1991).

The decision to house captive primates in social groups raises at least three questions: (1) What should be the size and age/sex class composition of the group? (2) How stable, in terms of introductions and removals of animals, should these groups be? (3) What should be the temporal patterning of this type of housing (i.e., intermittent or continuous)? Group size and structure are often the most difficult features of social organization to reproduce in laboratory settings. Although many primates naturally exist in large multimale, age-stratified social groups, space considerations often prevent the duplication of large troops in the laboratory. Two strategies may be employed in these situations. One involves creating groups that reflect subunits of larger troops (e.g., the harem unit of the hamadryas baboon) and the other involves forming groups that represent the smallest grouping observed in nature for a given species. For example, although rhesus monkeys generally live in relatively large troops (i.e., 25–200 individuals), they are sometimes found in small groups (i.e., 6–20 individuals) which can be simulated more readily in laboratory settings (Altmann and Muruthi, 1988).

Stability of membership is another important variable used to characterize social groups. Many primate species exist in relatively closed societies, and movement of animals between groups may occur primarily within specific age/sex classes. In rhesus monkeys, for example, troops mutually avoid one another or are overtly antagonistic during encounters, especially during breeding seasons. In addition, such troops are frequently xenophobic and respond aggressively to strangers (Southwick ct al., 1974). Females are the core of the group with mothers and their daughters forming sub-units known as matrilines. Although membership in these groups is quite stable, composition is altered as a result of births and deaths and the emigration/immigration of juvenile males (Colvin, 1986; Wrangham, 1980). These findings suggest that rhesus monkeys should be maintained in relatively stable groups although it may be important to remove juvenile males (i.e., re-create emigration). Given the lack of opportunity for animals to escape aggression in laboratory groups, attempts to introduce strange juvenile males into social groups can lead to severe injury (Bernstein et al., 1974).

Although we commonly think of social housing as the continuous maintenance of animals with social partners or group members, intermittent social housing, in some species, may more accurately reflect their social organization. Here, the word "intermittent" refers to regular social interactions that take place on a daily or weekly basis. Zoos frequently employ a type of intermittent social housing in which the same animals are displayed together during the day but may be housed in smaller groups or individually at night. This particular strategy has been employed successfully with orangutans, who, by their more solitary nature, may benefit from some time by themselves (Poole, 1987). Continuous housing would be considered more natural for troop-living species of primates such as squirrel monkeys, macaques, and baboons.

By maintaining captive primates in social groups, we may go further in creating a more natural environment than we do by merely altering the physical cage environment. Social stimulation has dynamic qualities. It can facilitate or impede other social interactions, it is seldom constant or completely predictable, and it is less likely to produce habituation than are most physical forms of stimulation. Social housing, however, includes certain negative features (Novak and Suomi, 1988). There may be an increased incidence of wounding, more rapid transmission of disease, and increased competition for food (all of which commonly occur under natural conditions). In addition, low ranking animals may be subjected to stress, tension, and physical attack on a regular basis.

THE RELATIONSHIP BETWEEN NATURALISTIC LABORATORY ENVIRONMENTS AND WELL-BEING

Naturalizing the laboratory environment is often considered to be synonomous with promoting the well-being of captive primates. Although this representation is usually accurate, certain factors such as the goal of the researcher, the process of simulation, and the background of the subjects may limit the attainment of well-being in primates housed in seminatural laboratory conditions. Any natural habitat consists of features that promote well-being (e.g., adequate food, nest sites, space to roam) as well as those that constitute deprivation or danger, and are therefore inimical to well-being (e.g., insufficient food or water supply, presence of predators, high population density). Depending upon the goal of the researcher, it may be important to simulate some of these detrimental, albeit natural, features in the laboratory in order to understand some basic ecological process (e.g., crowding studies as a basis for learning about population dynamics, response to predators).

In still other instances, the transposition process itself may be problematic. We may be unable to transpose the feature in an appropriately homologous or analogous manner, or the effect of isolating the feature and simulating it in the laboratory may be different than its cumulative effect in nature. Furthermore, the selection of specific features to simulate in the laboratory environment may have little relevance for the primates themselves. These problems with the simulation process may account, in part, for the paradoxical findings noted in previous sections (e.g., increases in cage space, use of partitions, and simulations of foraging activity can sometimes produce heightened aggression). Thus, we cannot always predict whether potential simulations will promote the well-being of captive primates. Both the benefit and generality of such simulations should be verified experimentally.

The extent to which seminatural environments elicit well-being may depend, in part, on the early rearing experiences and later housing history of the primates under study. Laboratory-housed, wild primates might be expected to respond positively when placed in environments engineered to resemble features of their native habitat. But there is debate as to whether laboratory-born primates would similarly benefit from these naturalistic simulations.

Recent evidence, however, clearly demonstrates that laboratory primates can and do benefit from exposure to naturalistic environments (O'Neill et al., 1991). In one study, monkeys were reared during the first year of life in socially restricted conditions in which they had continuous access to an inanimate surrogate and brief daily exposure to peers reared in the same manner. After the first year of life, the surrogates were removed and the monkeys were housed in their respective peer groups in wire mesh cages. At 21 months of life, eight of the monkeys were moved to a more natural outdoor environment which consisted of a corncrib containing numerous perches, swings, and toys. Four of the monkeys remained in the laboratory in their smaller, less complex, wire mesh cages. Four months later when the outdoor temperature became prohibitively cold, the "corncrib" monkeys were returned to the standard cages in the laboratory. Major differences in behavior were noted between the two groups. Corncrib monkeys showed a reduction in self-directed activity and passive social, and an increase in locomotion and exploration in comparison to the laboratory-housed monkeys (see Fig. 16.3).

Movement to a more naturalistic laboratory pen also appears to benefit surrogate-peer-reared rhesus monkeys. Two-year old monkeys that were placed in a large floor-to-ceiling pens containing shelves at various heights, swings, partitions, and a variety of manipulable objects (see Fig. 16.4) displayed little self-directed activity, and they showed increases in locomotion, exploration, and play in comparison to animals maintained in standard group caging. When these monkeys were subsequently maintained for a ten-year period in these enriched pens, they showed a remarkable range of species-normative behavior that would not have been predicted from their early rearing history (Novak et al., 1992). For example, basic sex differences emerged, with females showing

THE EFFECTS OF OUTDOOR CORN-CRIB HOUSING

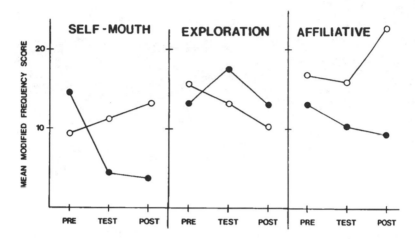

Figure 16.3. Behavioral change associated with the movement of surrogate-peer-reared monkeys to an outdoor corn crib pen.

more contact and males spending more time by themselves. Furthermore, females exhibited normal maternal behavior even though they themselves had never been exposed to a mother (see Fig. 16.5). Finally, after several birth seasons, the formation of matrilines could be observed as evidenced by higher contact scores between mothers and their daughters than between unrelated females (Novak et al., 1989).

The data cited above clearly indicate that many aspects of species-normative social organization and behavior will emerge in rhesus monkeys maintained in social groups, even if they have been reared under moderate conditions of social deprivation—that is, surrogate-peer rearing. What, if any, are the "limits" on the range of species-normative propensities and activities such monkeys might display if they were returned to a "real" natural habitat or if they were moved into a highly homologous outdoor environment? Two recent studies bear directly on this issue. The first, involving the reintroduction of captive-born golden lion tamarins to natural habitats in Brazil, is described in detail in chapter 17 by

Figure 16.4. Monkeys maintained in naturalistic outdoor enclosure at the NIH Animal Center. (Top) a view of the outdoor area; (Bottom) A closeup view of the corn crib shelters.

Beck and Castro in the present volume. The second study involved a long-term follow-up of the above-described surrogate-peer reared rhesus monkeys who were moved outdoors into a "corncrib" environment during their second year of life (Suomi and O'Neill, 1987). Over the next 9 years, this group of captive-born monkeys was moved into two different multiacre outdoor enclosures during the spring, summer, and fall of each year, returning to group cages in the laboratory during the winter months. They were subsequently moved into a third multiacre enclosure with a winterized shelter that enabled them to spend the entire time with access to the out-of-doors (Fig. 16.6). This surrogate-peer reared group of rhesus monkeys has been maintained without interruption in this highly naturalistic environment for the last five years. During all this time, the monkeys were provisioned daily and were never exposed

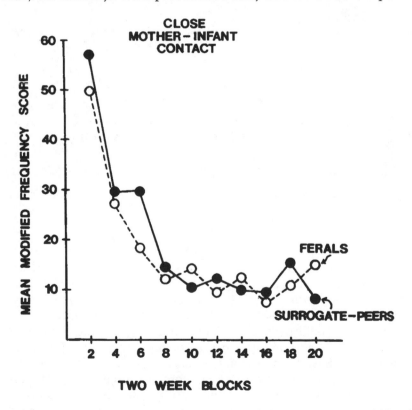

Figure 16.5. A comparison of close mother-infant contact in wild-born, laboratory-housed monkeys and lab-born (surrogate-peer-reared monkeys) in an enriched indoor pen.

to mammalian predators or to other groups of conspecifics. Nevertheless, they did have unrestricted access to a large outdoor area containing many species of trees, forbs, and grasses, as well as a large pond, that supported many native species of wildlife.

Figure 16.6.   One section of an enriched indoor pen with multiple perches at various levels.

Extensive long-term observations clearly demonstrated a remarkable capacity of these monkeys to develop species-normative patterns of behavior and social organization when moved into highly naturalistic outdoor settings. The original group of eight juveniles has grown to over 22 individuals spanning three generations, despite some losses by death and by forced emigration of virtually all the adolescent males. The "founding generation" of juveniles developed appropriate patterns of reproductive and maternal behavior. Both first- and second-generation offspring have, in turn, displayed normal sequences of behavioral development from infancy to adulthood. Moreover, as the number of individuals and generations within the groups has grown over the years, a clear pattern of social organization based on multigenerational matrilines has emerged, with complex dominance hierarchies that appear to be similar to those reported for wild populations studied in India, Nepal, and Cayo Santiago (Chepko-Sade and Sade, 1979; Missakian, 1972; Southwick et al., 1965). In this regard, they are also similar to surrogate-peer-reared monkeys raised in complex indoor group cages (Novak et al., 1992).

Exposure to outdoor settings, however, has enabled these captive-born monkeys to exhibit capabilities not readily apparent in indoor environments. For example, upon moving outdoors year-round, these monkeys quickly developed seasonal patterns of behavioral change mirroring that seen in wild troops, including the emergence of well-defined breeding and birth seasons. In addition, they all learned to supplement their provisioned diet with grasses, shoots, berries, nuts, insects, and trace minerals (obtained through geophagy) in each of the outdoor settings to which they were exposed. Moreover, the extensive space available to these monkeys permitted the development of species-normative social dynamics—dominance relationships between multiple matrilines or changes in affiliative patterns and space utilization leading to adolescent male emigration, that would likely be masked in indoor environments (Suomi and O'Neill, 1987).

The discovery that laboratory-reared rhesus monkeys readily adapt to naturalistic habitats and retain species-normative behavioral repertoires and patterns of social organization in the absence of any exposure to wild-born or socially sophisticated adult conspecifics is of considerable significance. It serves to demonstrate that complex behavioral capabilities are preserved in the genetic heritage of members of an advanced nonhuman primate species, even after many generations of laboratory living. Indeed, the relative ease, speed, and thoroughness of adaptation to a natural environment shown by these rhesus monkeys should encourage primate conservation efforts based on the establishment of captive breeding colonies with subsequent introduction into protected naturalistic habitats

SUMMARY

A major purpose of creating indoor naturalistic habitats is to promote species-normative patterns of behavior and social organization in laboratory primates. This is important not only for researchers who focus on behavior, but also for biomedical scientists who would benefit from examining the effects of physiological manipulations in an environment in which a broad range of responses is possible.

The creation of naturalistic environments involves transposition by homology, a direct duplication of some environmental structure, or transposition by analogy. In the analogous case, certain ecological challenges are re-created, and the emphasis is placed on biobehavioral outcome and not on the structure used to produce it. Increasing the complexity of the cage environment with perches, poles, and partitions; introducing the challenges of foraging and tool using; and providing social companionship are clearly effective in broadening the range of species-normative behavior shown by primates in captivity. Moreover, by housing primates in naturalistic settings, we take definitive steps toward promoting their health and well-being. This effect is not limited to wild primates who are now maintained in captivity, but extends to primates born and then raised in a variety of ways in the laboratory. Even rhesus monkeys reared with limited social experience show a remarkable ability to adapt to seminatural and natural settings and to exhibit complex patterns of social organization, which are characteristic of free-ranging monkeys.

BENJAMIN B. BECK
MARIA INÊS CASTRO

17

Environments for Endangered Primates

THE FALL OF THE WILD

We tend to think dichotomously about animal environments, contrasting naturalistic with nonnaturalistic, free-ranging with enclosed, indoor with outdoor, and wild with captive. Our work with the endangered golden lion tamarin (*Leontopithecus r. rosalia*) has led us to question these dichotomies. Animal environments instead fall on a continuum (Beck, 1980) from that in which a species evolves—for example the eastern coastal rainforest of Brazil for the golden lion tamarin—to small restrictive enclosures that allow control of both primate and extraneous variables for experimentation.

The "wild" for golden lion tamarins is currently the 5000-hectare Poço das Antas Biological Reserve, as well as a nearby restricted military base and a handful of forest patches on private land in the wild. Our colleague James Dietz, who is conducting a behavioral ecology field study, estimates that about 400 tamarins remain in the wild. The region is heavily populated and is being developed rapidly for cattle ranches and citrus plantations. Cut trees arc used for factory and household fuel. No surviving tamarin lives out of earshot of chainsaw and bulldozer. We have rescued tamarins huddled in terror under smoldering brush piles in freshly logged and burned clearings.

The Poço das Antas Reserve is a mosaic of secondary forest and old field. It is bisected by a railway and flanked by a highway. A new dam is altering the course and flood pattern of a bordering river. Fire periodically ravishes the fields and recolonizing tree seedlings; in February 1990 a fire burned over 30 percent of the reserve. Fighting fires and building fire breaks are prominent conservation activities. Gates and some fencing bar access at the most convenient entrances, but poachers, recreational motorcyclists, and squatters are common. The reserve's guard corp is too poorly manned, equipped, paid, and trained to provide total protection.

The behavioral ecology study began in 1983 as part of the Golden Lion Tamarin Conservation Program (Kleiman et al., 1986). Dietz has attempted to trap every wild tamarin. Each animal is brought to project headquarters and chemically anesthetized. It is weighed, measured, sampled for ecto-(parasites) and endoparasites, tattooed, and given a preventive dose of antibiotics. There are thirty-five to forty tamarin groups in the reserve, of which twenty are under current study (Baker et al., in press). Most groups have only one breeding male and one breeding female (Baker et al., in press); mean group size is about 7 (Peres et al., in press). Some animals in each group are radio-collared. Blood for genetic and virological studies has been drawn from many tamarins, and others have been radiographed. Tamarins with injuries have been medically treated. The groups are retrapped annually (the animals are so hungry that they will reenter traps for a bit of banana). Rescued or confiscated wild-born tamarins have been translocated into the Reserve and captive-born tamarins have been reintroduced. We have even discussed "reintroducing" gametes from zoo tamarins or "translocating" embryos to bolster genetic diversity.

LouAnn Dietz has conducted a conservation education project as part of the tamarin program. Most of this crucial work is conducted outside the Reserve, but classes are regularly brought in to see the tamarins and the forest from a special nature trail. The trail is actually a bit of a large trail system cut carefully through the Reserve for researchers and guards. This slight destruction is balanced by forest regeneration due to fire control, application of fertilizers, reduced cutting, planting of native tree seedlings, and translocation of bromeliads, but the trail system is clearly not natural. Even chartered tourist groups are occasionally allowed on the trail to see wild tamarins in exchange for a contribution that is used for Reserve management.

Poço das Antas is still a wild place, with rare orchids, brightly colored birds, coral snakes, and ocelots. But it is not pristine. The drone of cars, trucks, trains, and planes rivals that heard around any urban zoo. Animal traps, anesthetic drugs, antibiotics, fecal samples, tattoos, radiographs, blood samples, fire control, locked gates, trails, signs, and tours, all tools of the zoo manager, are essential for the continued survival of the golden lion tamarin. And if the tamarin can be saved, so too will be the less endearing species at lower trophic levels. That is, if the tamarin is saved, its environment is saved as well. But is it a naturalistic or nonnaturalistic environment? Are the tamarins free-ranging or enclosed? Are they wild or captive? These questions pertain not just to Poço das Antas and golden lion tamarins. Every national park is becoming increasingly insularized and penetrated by human influence that far exceeds the long-evolved subsistence uses that were sustainable only a century or two ago (Altmann and Muruthi, 1988).

CAPTIVE ENVIRONMENTS

The continuum of tamarin environments is being compressed from the other end as well. We know of no biomedical or behavioral program that keeps a golden lion tamarin for long periods in a small cage or restraining chair. Such settings may be necessary for essential research using nonthreatened primates, although one could argue that there are no nonthreatened primates or that scientific progress never justifies such loss of reproductive opportunity or the stresses of physical confinement and social deprivation. Some endangered primates are designated as surplus to captive breeding populations because of genetic overrepresentation or sex-ratio inbalances. But to use these individuals in invasive research would discourage the search for and use of management techniques that would preclude surplus animals.

But some conservation-driven work by our colleagues David Wildt and Steve Monfort does require temporary housing of golden lion tamarins in small metabolic cages for urine collection. Wildt and Monfort track urinary hormonal metabolites to noninvasively pinpoint ovulation. The data will coordinate artificial insemination and embryo transfer should these technologies be needed in zoos or in the wild for tamarin propagation. The results also provide physiological measures of sexual receptivity. Following the work of French and Stribley (1987), Wildt and Monfort study receptivity when a female lives alone, with other females in the absence of a male, with only a male, and with other females and a male. This provides better understanding of the mating systems of wild tamarins, and may reveal possibilities for increasing reproductive output. The subjects in this research live in conventional cages but are trained to enter the metabolism cages. The National Zoo's animal care and use committee determined that the potential yield of scientific information important for the conservation of the species justified the temporary separation of a few individuals.

The Golden Lion Tamarin Management Plan, a model for later plans for other endangered species, stipulates a cage size of at least 3 x 2 x 2.5 meters. That is roughly the size of a shower stall, but it is in such cages that the captive golden lion tamarin population has grown from about 70 in 1973 to over 550 today. The key was not the size of the physical environment but the suitability of the social environment. Kleiman et al. (1986) discovered that this species breeds best in captivity when kept as monogamous pairs, and when offspring are allowed to remain with their parents to assist in the raising of subsequent litters. Nest boxes, climbing networks, elevated feeding stations, and the addition of meat protein to the diet also contributed by simulating or substituting for essential resources of the natural environment. But the research and breeding that made the propagation program for golden lion tamarins so successful took place in small, inexpensive, easily maintained, notably unaesthetic cages in more than 50 zoos around the world.

Some zoos are now housing tamarins in large naturalistic habitats. Zoo habitats tend to be stage sets (Robinson, 1988) with artful rock work, artificial trees, vines, lichens, pools, waterfalls, rainstorms, theatrical lighting, and auditory surrounds. Live vegetation can be included with tamarins since they do not eat leaves (although locomotion will take a toll). South American rodents and sloths are appropriate cagemates. But habitats do little for the tamarin within. Their strength is that they delight and comfort the zoo visitor outside, who hopefully becomes more sympathetic toward and knowledgable about the preservation of biological diversity.

PREPARING TAMARINS FOR REINTRODUCTION

We thought habitats would be better than cages for tamarins being prepared for reintroduction. Foraging skills could be facilitated because food could be distributed over a greater area and there would be a greater range of places to hide food. More important would be the greater locomotor opportunities afforded in a habitat than in a cage. Captive-born, cage-housed tamarins are reluctant to use slender flexible vegetation; indeed if their cage contains only a rigid climbing structure of milled lumber or plastic pipe, they are reluctant to use natural vegetation of any sort. We hypothesized that a habitat outfitted with vines and slender branches as the only means to reach nest boxes, resting sites, food, and water would predispose tamarins to acquire the locomotor skills essential for survival in the wild. Further, because tamarins living in cages with unchanging climbing structures develop habitual travel routes and are unable to spatially orient after reintroduction, we thought that dismantling and rearranging the naturalistic furniture in the habitat would facilitate spatial orientation. Training in locomotion and orientation was thought to be especially crucial, since a tamarin deficient in these skills would be seriously disadvantaged in finding food and shelter, avoiding predators, and maintaining normal intragroup and intergroup social behavior (Beck et al., in press; Fleagle, 1979; Garber, 1984). But the labor-intensive pretraining in a habitat conferred only short-lived postreintroduction advantages in foraging, locomotion, and spatial orientation (Beck et al., in press).

Further, ingenuity can provide equivalent experiences in small cages. For example, tamarin cages in Snowdon's laboratory are outfitted with "rubber perches," which are excellent simulations of slender natural vegetation (Snowdon and Savage, 1989). McGrew and McLuckie's (1986) flexible ducts allow a tamarin to leave its cage and explore a large area and neighboring tamarins. This facilitates orienting skills and provides opportunities for the animals to learn the "etiquette" of territorial encounters.

We know of no evidence that large, naturalistic habitats enhance the well-being or survival skills of tamarins. Functional (not necessarily naturalistic or aesthetic) simulation of critical features (or salient elements; see Glickman and Caldwell, chap. 14, this volume) of the natural physical environment (Fig. 17.1), and a natural social environment, are sufficient for health, longevity and reproduction, and for the expression of a broad sample of the behavioral repertoire. Further, functional simulation conserves the scarce resources available for captive breeding and research programs. Tasks requiring the tamarins to work for food, and thereby gain some control over feeding schedules, seem also to contribute to well-being (Kleiman et al., 1986; Markowitz, 1982; Molzen and French, 1989; Snowdon and Savage, 1989).

Figure 17.1.   A klipspringer (*Oreotragus oreotragus*) on a rocky outcrop in east Africa. Photo by B. Beck.

We built habitats at the release sites of the golden lion tamarins that we reintroduced to the wild in 1984 and 1985. These habitats were wood and wire mesh cages, 15 meters long, 4.5 meters wide, and 3 meters high. The understory of the forest was bent in and down as the mesh was assembled around it; taller trees extended through the mesh at the top. Each habitat had a nest box. These habitats were built to allow the animals to acclimate to and become familiar with the release site, gain some locomotor and foraging experience in natural vegetation, encounter invertebrates and small vertebrates that came to the cage, and experience direct rainfall. Further, five of the six groups reintroduced in these years had been intensively trained in foraging and locomotion before being moved to these forest cages, where their training continued. In retrospect these habitats were a forgiveable conservatism, but

Figure 17.2. Klipspringers on rocks in the Frankfurt (Germany) Zoo. Though the enclosure is decidedly nonnaturalistic, the rocks functionally simulate outcrops, a critical habitat element for this species. Photo by B. Beck.

they did not serve their purpose. For reasons noted above, the tamarins, especially the adults, preferred to use the wood framing and wire mesh for locomotion. We documented that the animals recognized knotholes and rolled leaves as likely placcs to search for food, and that they would eat spiders, butterflies, and snakes if they could catch them. They seemed indifferent to rain, unless they had full bellies in which case they went into their nest box. But we do not think the tamarins actually learned much in these habitats; they clearly could not have survived in the forest without extensive human support over at least a year.

The moment of reintroduction was marked by opening a small door in the roof of the habitat. Fleet-footed field assistants with radio receivers were deployed around the area to follow the tamarins as they rushed back to nature. But the tamarins seemed reluctant to leave the habitat, venturing only 30 meters away and 20 meters high on the first day. They played on and in the 3-meter high habitat during the day, and slept inside in the nest box at night. Because the understory had been compressed into the habitat, tamarins on the roof were dangerously exposed to raptors. We had to dismantle the habitats shortly after release.

One of the groups in the 1985 release received no formal training, although it did live for an equal time in the forest habitat before actual reintroduction. As noted abovc, this untrained group showed poorer food-finding and ranging after reintroduction, compared to a matched trained group, but this difference was only transient. After documenting the difference, we felt obligated to train the untrained group. This was of course possible since the animals had remained near the prerelease habitat site. Food was hidden in slots in a special platform, and the platform was moved farther and farther into the forest each day. The animals learned to search for food in the platform, and in natural sites that they encountered on their way to and from the platform. They also learned to locomote and orient in and through real natural vegetation. This seemed to be a more authentic method of training and it eliminated the need for costly "habitats in the wild" that were of limited value. We fundamentally revised our reintroduction strategy to make use of a new type of tamarin environment.

THE PSYCHOLOGICAL CAGE

The psychological cage depends less on containment than on attachment. A captive-born group of tamarins with a nest box and a food supply, and without strange tamarins calling in the distance, is not motivated to move when released into the forest. Lack of locomotion and orientation skills compound the sedantism. Harking back to environmental dichotomies, they are surely "free-ranging" but they are actually "confined." We reasoned that tamarins could be more realistically and cost-effectively trained after release than before. We could do this safely since they would be in their psychological cage. Wim Mager, director of the Apenheul Zoo, independently converged on this idea, and uses it to mount unique primate exhibits (Mager and Griede, 1986). Several callitrichids and squirrel monkeys roam in their psychological cages at Apenheul. We set out to develop the concept by releasing a tamarin group in a fine stand of beech and oak forest in the heart of the National Zoo (Bronikowski et al., 1989). The zoo's animal care and use committee had a hard look at this proposal as well; letting animals out of cages is heretical to zoofolk.

We caught the tamarin group before it emerged from the nest box into its colony cage on the morning of the release. The box was taken to the release site, mounted in a tree, and the entrance opened. The feeding platform was hung just outside the entrance. The monkeys made brief cautious forays but were predictably reluctant to stray. On the second day they ventured farther, but then could not easily find their way back to the nest box. Some returned on the ground. Some had to be lured back with banana. Others fled and had to be trapped and carried back to the nest box tree; one was lost. Subsequently, none of the tamarins ventured farther than 100 meters from the nest box for the next three months (when the release was terminated due to cold weather). We repeated the experiment the following spring with the same group.

We strung stout ropes horizontally between mature trees in the immediate release area to compensate for the lack of the understory that wild tamarins seem to prefer. Food was again available in abundance. The animals did not "escape" from their psychological cage, but rather gradually expanded its edges. They visited trash cans and pop stands, and invaded the enclosures of bush dogs (which cowered), eagles (which cowered), bison (which charged), oryx (which seemed perplexed), and gibbons (which bit two tamarins severely, killing one). They had regular morning territorial encounters with another golden lion tamarin group in an outside habitat 200 meters away. They raised twins. The tamarins in this environment delighted and educated zoo visitors.

We were convinced that the psychological cage could be used to train reintroduced tamarins since we had learned that fully provisioned tamarins would slowly but surely begin to explore their environment, and in the process would learn to hunt effectively for "natural" food and to locomote and orient in "natural" vegetation. Jim Doherty (personal communication), General Curator, New York Zoological Society, has released ring-tailed lemurs in psychological cages on St. Catherines's Island with similar results. And we have subsequently used postrelease training in psychological cages in the reintroduction of more than 50 tamarins to the wild in Brazil. Although as we predicted, some animals initially fled and had to be recovered, they later expanded their territories around their nest boxes, had territorial encounters with neighbors, and reproduced. The notion of psychological cages holding reintroduced tamarins in Brazilian forest obliterates environmental dichotomies such as "captive" and "wild." That they willingly return to their nest box, and spend more than 12 hours per day inside the box in less than a cubic meter of space, obliterates the dichotomy between an "indoor" and "outdoor" environment, and tests our biases about minimum cage size standards.

NATURALISM, STRESS, AND WELL-BEING

Even wild-born tamarins living totally at liberty in the Poço das Antas Reserve are not really free: "However paradoxical it may sound, the truth is actually this: the free animal does not live in freedom" (Hediger, 1964, p. 4). They engage in energy-consuming territorial encounters with neighboring groups (Peres, 1989), demarking for us researchers the unseen boundaries that surely constrain each group in a slightly different type of psychological cage. They roam their 40 hectare territories seeking the cryptic food that is distributed sparsely in space and time. Predators such as hawks, ocelots, and boas stalk the unwary. Nights are wet and surprisingly cold, and good nest holes are hard to find because there are few really old trees. Mosquitoes, ticks, and fly larvae feed on tamarins too. This hardly conforms to the romantic notion of the natural environment as a paradise, and serves as a reference for some thoughts on animal welfare, "psychological well-being," and stress.

Moberg (1985b) argues that stress can be measured in nonverbal animals by its prevention of normal reproduction, alteration of metabolism so as to cause weight loss and interference with normal growth, suppression of the immune system so as to increase the likelihood of infection-related death, and hyperaggression or self-mutilation. If these effects can be observed in a captive animal, then that animal can be said to be stressed. The average captive-born adult golden lion tamarin, even one living in a shower stall, weighs 675 grams. The average wild adult weighs 600 grams (J. Dietz, personal communication). Captives are bigger as well as heavier. Captivity is certainly not interfering with growth or causing weight loss. Captive golden lion tamarins typically give birth to two litters per year; wild tamarins typically give birth to one. There is certainly no interference with reproduction. Hyperaggression and self-mutilation among captives disappeared after Kleiman's discoveries about keeping golden lion tamarins in monogamous pairs and allowing offspring to gain parental competence by helping to raise their younger siblings. Finally, the average longevity of zoo tamarins is now about 12 years. One has survived beyond 20! Dietz is not yet able to estimate life expectancies for wild tamarins, but their average longevity is certainly less than twelve. Further, preliminary data show that zoo-born reintroduced tamarins lose weight and have a birth rate like that of wild tamarins.

These comparisons suggest that "the wild" is far more stressful, and captivity (under good but not extravagant conditions) is far less stressful, than the more radical animal welfarists would have us believe. Recall that in our reintroduction program we gave golden lion tamarins the choice of being wild or captive: they chose to remain captive. Of course they may have chosen captivity out of familiarity, but remaining in captivity would nonetheless have been safer: at this time only 32 of 79 reintroduced tamarins (and 22 of their wild-born offspring) are still alive. Viral disease, exposure, intraspecific conflict, toxic fruits, predators, and theft by humans are documented or implicated as causes of loss. We would need to know age-specific mortalities of wild tamarins to compare this survival rate with that of a like-aged cohort of wild tamarins over a comparable period, and a like-aged cohort of wild-born tamarins brought into captivity. But we do know that if we had brought 79 wild tamarins into captivity since 1984, far more than 32 would be alive today. We must inescapably conclude that tamarin well-being is enhanced by good captive management; that stress is diminished in captivity; and that naturalistic environments do not necessarily promote health and reproduction. Indeed, effective preparation of captive primates for reintroduction requires environments that are more naturalistic—that is, that provide more stress and less well being (Beck et al., in press; Kleiman et al., 1991).

ACKNOWLEDGMENTS

The Wildlife Preservation Trust, Friends of the National Zoo, Frankfurt Zoological Society—Help for Threatened Wildlife, and a number of participating zoos have provided funding specifically for the reintroduction of golden lion tamarins. M. Inês Castro's graduate study was supported by a CNPq-Brazil scholarship, and her travel to the Conference on Indoor Naturalistic Facilities was funded by the Wildlife Preservation Trust. Staff of the department of mammalogy and the department of animal health at the National Zoo made the free-ranging exhibit work. Many dedicated volunteers from the Friends of the National Zoo helped to monitor the free-ranging tamarins. Mike Robinson, Devra Kleiman, Jim Dietz, and Mike Power discussed many of our ideas with us and critically reviewed the manuscript. We are grateful to all.

CHARLES H. JANSON

18

Naturalistic Environments in Captivity: A Methodological Bridge between Field and Laboratory Studies of Primates

Studies of animal behavior in the laboratory have traditionally focused on mechanisms of behavior: the neural, physiological, developmental, and psychological processes that give rise to behavior we observe in a given species. Conversely, studies of animal behavior in the field have usually emphasized evolutionary questions: understanding the natural context within which behaviors develop and are expressed, and the important forces that may have shaped those behaviors via natural selection (Tinbergen, 1963). Any complete understanding of the behavior of a given organism must encompass both approaches. In addition, however, it is argued here that a pure reliance on field studies to answer evolutionary questions and on laboratory studies to answer mechanistic questions in behavior necessarily leads to an incomplete understanding of either type of question. Laboratory and field approaches can complement each other in specific ways to provide more comprehensive answers to both evolutionary and mechanistic problems. The examples in this paper are mostly from primates, both because this is the group I am most familiar with, and because it is one of the most intensively studied groups of animals in captivity.

HOW CAN LABORATORY STUDIES PROVIDE EVOLUTIONARY INSIGHTS?

Most evolutionary field studies on animal behavior involve gathering information for a set of known individuals: (1) a detailed qualitative and quantitative description of the behavior(s) of interest, (2) individual phenotypic or ecological characteristics, and (3) some measure of individual fitness. Leaving aside the question of whether or not this approach alone is adequate to study evolutionary aspects of behavior, laboratory studies can act as useful adjuncts to this research program.

First, many social behaviors are too infrequent or subtle to be recorded easily in the wild, even if observation conditions are excellent (which they rarely are). Thus, a laboratory study may help to describe the behaviors used by a species, and allow less conspicuous behaviors to be recognized in the field by their particular ecological or behavioral contexts (Moynihan, 1964; Eisenberg, 1976; de Waal et al., 1976; de Waal and Yoshihara, 1983). As an example from my own work, despite observing some 50 heterosexual copulations in brown capuchin monkeys in the field (Janson, 1984), I was unable to determine the contributions of male vs. female to copulatory vocalizations, a dilemma that does not confront researchers on primates in captivity (Hamilton and Arrowood, 1978) or seminaturalistic environments (Doyle et al.,1967).

Second, our senses are incapable of perceiving many phenotypic traits of an organism that may be critical to explaining the occurrence of a behavior. For instance, the timing of ovulation in primates is crucial to determining male or female reproductive success, yet until recently there were no field methods available to determine the precise time of ovulation. In this particular case, many advances in field work can be traced to laboratory studies linking physical and hormonal changes (e.g., sexual swellings in baboons, as used by Hausfater, 1975) or providing improved methods of detecting hormonal state noninvasively (e.g., as used by

Andelman, 1985). Similarly, knowing the kin relationships of an individual can be important to interpreting its behavior, yet at best a field worker using observation can be sure only of maternal relationships, while in the laboratory, relationships can be manipulated by breeding designs, or assessed using genetic markers (Duvall et al., 1976). Recent advances in molecular genetics promise similar ability to resolve kinship relations even to field workers using noninvasive methods (Westneat, 1990).

Third, the fitness consequences of behaviors are frequently difficult to evaluate in the field, because most behaviors affect only some correlate of fitness (e.g., food intake), and the relationship between the latter and total fitness is usually unknown. Laboratory experiments can be designed to measure precisely the fitness effects of given behavioral or ecological manipulations. For example, captive studies of energy expenditure have provided data important in estimating the energy budgets of wild primates (Janson, 1988b). Other studies of primate nutrition have assessed the impact of various levels of protein versus calorie intake on fetal health (Riopelle et al., 1975) and on juvenile growth (Samonds and Hegsted, 1978), both of which contribute to total fitness. Similarly, when several males copulate with a receptive female, it would be interesting to know how copulation order affects a male's chance of siring offspring (as has been determined in laboratory studies of certain insects, e.g., Waage, 1979).

The study program described at the start of this section is largely inductive: the researcher first uncovers a set of basic facts about a species before these are related to more general principles governing behavior. This descriptive approach is often criticized as unscientific (Pierce and Ollason, 1987), because hypotheses are formulated only after the data are collected. To counter these criticisms, there is increasing emphasis on testing these post-hoc hypotheses with experiments. Whereas certain experiments can be performed in the field, there is always the danger that the manipulation performed is perceived incorrectly by the animal, or that the manipulation becomes confounded with other changes in the animal's natural or social environment.

The control possible in the laboratory allows rigorous testing of the importance of putative causal agents in affecting behaviors. A successful example of this approach is that of captive vervet monkey alarm-calling behavior (Cheney and Seyfarth, 1985), which demonstrated that females call more frequently when relatives are present than when they are absent, a difference that could not be evaluated in the field (Cheney and Seyfarth, 1981). As

another example, metabolic studies on captive primates have provided crucial data to test the hypothesis that evolution favors
more efficient forms of locomotion within a taxon (Fleagle, 1979).

There are many similar questions that remain virtually unex-
plored. For example, a long-standing question in behavioral ecology is how food abundance and distribution interact with group size
to affect individual food intake (Janson, 1988a). Although certain
insights are possible from field studies (Chapman, 1988; Janson,
1985, 1988b; Symington, 1988), these insights are necessarily
based on correlations, which can mask or confound true causal
relationships. Well-designed laboratory studies manipulating food
availability and social structure while monitoring resultant individual food intake would go far in distinguishing direct causal relationships from indirect correlated responses. It is unfortunate that
few such studies have been performed (Belzung and Anderson,
1986). In addition to revealing causal effects, such studies could
demonstrate the behavioral mechanisms by which ecological variation is translated into variation in food intake. Knowing these
behavioral signs would vastly increase the ability of field
researchers to describe and predict competitive relationships in
nature.

A second example concerns food choice in primates. Data
from the field on whether primates choose foods in an optimal
way are varied, and much controversy still exists (Glander, 1978;
Milton, 1980; Post, 1984). Observational studies on primate food
choice suffer from several obvious problems: unequal availability
of different food types (Janson et al., 1986), social interactions (Post
et al., 1980), and difficulty in collecting precisely what the animals
are eating (Malenky, 1990). Laboratory experiments offering isolated individual primates equal access to various foodstuffs of known
nutrient value could easily test hypotheses about foraging rules,
yet this approach is rarely used (but see Glander and Rabin, 1983;
Milton, 1984). This dearth of studies is probably caused in large
part by the usually divergent study interests of field versus laboratory primatologists (Rosenblum et al., 1989).

The preceding questions about food competition and food
choice would benefit greatly from research performed in seminaturalistic environments. One of the strongest constraints on primate
foraging success in the wild is the need to find food sources of
unknown location (Janson, 1990). Typical laboratory settings are
rarely large enough or complex enough to present a primate with
such foraging challenges. Thus, only questions about behaviors fol-

lowing food discovery can be adequately addressed in typical laboratory cages. Seminaturalistic environments can provide most of the benefits of ease of observation and experimental control obtained with standard cages, yet offer sufficient complexity to allow study of the 75% or more of an animal's behavior that occurs when looking for, rather than exploiting, a food source (Menzel, 1973; Menzel and Juno, 1985).

The research approach introduced at the start of this section implicitly assumes that behavior is or can be optimized, and thus that the common behaviors used by individuals of a species are generally those that produce the highest fitness from among the available alternatives (Vehrencamp and Bradbury, 1984). However, recent evolutionary controversies have highlighted the importance of various kinds of constraints in shaping the traits of extant organisms (Gould and Lewontin, 1978; Maynard Smith et al., 1985). It may be less interesting to show that a given behavior is the best available than to ask why other, even more fitness-enhancing behaviors, are absent in the species examined. The latter question often requires a detailed knowledge of development, genetics, and evolutionary history that cannot be acquired purely by field studies.

Development is likely to be important whenever behaviors are primarily learned. It may often occur that an individual fails to perform a given behavior because it never had the opportunity to learn the form or context normally associated with that behavior. For instance, whereas alarm calls in primates are often quite constant in form within a species and thus probably not learned (Gauthier, 1988), it is likely that primates are not born knowing the object of specific alarm calls (Seyfarth and Cheney, 1986). If so, then infant primates must learn what animals are dangerous by observing the behaviors of their parents. It is then easy to imagine that local "traditions" of alarm reactions could develop, including individuals or groups that fail to show alarm responses to novel predators (including humans). Similarly, there is now strong evidence that many aspects of foraging behavior require observational learning by juveniles of some primates (e.g., golden lion tamarins; Kleiman, 1989), although not in other species (squirrel monkeys; Boinski and Fragaszy, 1989).

Probably the broadest class of unknown constraints in behavioral studies are genetic. Although some behavioral differences within species are known to have a simple genetic basis (see Fuller and Thompson, 1960), most behaviors are the end product of a necessary interaction of genotype and environment, which cannot readily be examined in the field. A first step to understanding this interaction is to examine behavior either (1) in standard genetic stocks under different developmental conditions, or (2) in different genetic stocks under identical conditions, as is possible in the laboratory. An example of the first approach is the obvious difference in socialization of isolate versus group-reared juveniles of a given species (Swartz and Rosenblum, 1981), which should be important in explaining or predicting behavioral differences that may arise in wild primates as a function of stochastic variation in numbers of infants born into a group in a given year (Altmann and Altmann, 1979). Similar interesting insights can be gained by behavioral differences that arise following cross-fostering of embryos of one species to another (usually closely related) species (Gibbons and Durrant, 1987). Assuming that the fostering process itself has no effect on subsequent behavioral development, such cross-fostering techniques are an ideal method to evaluate the role of genetic versus environmental factors in accounting for differences in behavior among related species. An example of the second approach is the comparative studies of social behavior of different primate species held under nearly identical conditions of group size and available space (Rosenblum and Kaufman, 1967; Mason, 1974; de Waal and Luttrell, 1989). Despite considerable variation within species (de Waal and Luttrell, 1989), behavioral differences between species found in these studies suggest that similar differences observed in the field have a genetic basis and are not just the result of the different ecological circumstances each species occupies as a result of its morphology and habitat. A more quantitative understanding of the genetic basis of behavior can be obtained from controlled captive breeding programs, which allow a comparison of behavioral similarity between parents and offspring or within versus between sibships (Ayres and Arnold, 1983; van Berkum and Tsuji, 1987).

Finally, captive studies can be used to answer questions about behaviors in contexts that never occur in the natural environment of a given species. Although such novel captive behaviors are often dismissed by field researchers as "abnormal," they can in fact provide interesting insights into the behavioral capacities of a species (as argued for positional and locomotor behaviors by Prost, 1965). Indeed, one can argue that such captive experiments reveal hidden potential for evolutionary change ("preadaptations"), which could not be observed in the field. One possible example is the use of tools by captive capuchin monkeys. Wild capuchins have been seen using sticks as "weapons" against threatening animals (usually human observers), but they have not (yet) been seen to make tools to obtain food, although they will do so readily in captivity (Fragaszy and Visalberghi, 1989; Visalberghi, 1988). It is possible that this behavior does occur in nature but is so rare it has been missed, or it may represent a presently nonadaptive side effect of a high cognitive intelligence, which evolved for other reasons. As the latter statement implies, thorough field studies are necessary to interpret if and how observed behaviors in captivity relate to the species' current selection pressures in the wild.

HOW CAN FIELD STUDIES PROVIDE MECHANISTIC INSIGHTS?

Much of the research effort in laboratory studies of behavior addresses questions of mechanism: psychological, neurological, or developmental. The essential approach in such studies is to manipulate the environment of the animal in ways appropriate to the hypothesis of interest and observe the resulting behavioral changes. Because field studies rarely have the control over environment or excellent observation conditions possible in captivity, it seems unlikely that field studies will provide critical tests of behavioral mechanisms beyond what is possible in the laboratory. However, field studies can both inform and inspire mechanistic studies in captivity.

The informative role of field studies to captive research may lie in the interpretation and understanding of behavioral variation. For instance, it is well-known that different captive studies apparently addressing the same mechanistic question can yield quite conflicting results (e.g., for the role of mother in infant development of independence, see Suomi, 1976). Much of this variation is probably due to unanalyzed differences between studies in variables not originally recognized as being important to the study question. In addition, different species often possess unique behavioral abilities or deficiencies that make it difficult to generate universal models of behavioral mechanism. An example is the very different capabilities in diverse organisms to store and relocate food items (Sherry, 1985).

In both the above cases, results from field studies may provide important insights. Just as a complete understanding of the evolution of behavior requires an understanding of the mechanisms by which behaviors come to be expressed, so too a proper understanding of behavioral mechanisms may require a knowledge of the social or environmental context within which these behavioral mechanisms function (usually) to the individual's benefit (Lorenz, 1965). For instance, the striking differences between squirrel monkeys (*Saimiri*) and titi monkeys (*Callicebus*) in spacing and social behaviors in captivity are not interpretable without knowing that titi monkeys in nature are monogamous and squirrel monkeys are highly polygamous (Mason, 1974). Similarly, knowledge of natural social structure may be important to understanding species-typical responses to social stress (Kaplan, 1986), or why filial and sexual imprinting occur at different times in development (Bateson, 1981). Insights about patterns of behavioral development are not available only from captive work. In general, it is not possible in captivity to assess accurately the intensity of natural selection experienced by animals in the wild. Thus, field studies can provide data important to interpreting observed patterns of captive behavior.

Finally, field studies can and do provide a wealth of unsolved behavioral questions, many of which are most appropriately answered by controlled captive experiments. Although laboratory behavioral research on primates has tended to focus on questions of behavioral development, some of the examples provided in this chapter show the potential for captive research to inform and test ecological and evolutionary hypotheses (as is done more routinely with insects, e.g., Waage, 1979). Although I have emphasized the importance of studies of captive behavior, there are also many physiological questions that are of great interest for evolutionary studies of behavior. To mention just one example, some primate populations are known to have markedly biased sex ratios at or close to birth (Clark, 1978b; Clutton-Brock, 1982). Although it is not known if such variation is adaptive, it would be very interesting (to humans) and informative (evolutionarily) to discover the hormonal or other mechanisms by which such variation is produced.

CONCLUSIONS

Laboratory studies can be important in sharpening the inferences from traditional adaptationist field studies of primate behavior. Laboratory studies can (1) provide details of behavior not easily observed in the field, (2) yield data to allow conversion of fitness correlates to actual fitness differences, (3) test hypotheses derived from field work, (4) provide data on developmental and genetic contributions to behavior under standardized conditions, (5) reveal latent behavioral capacities not demonstrated in nature. Although not all laboratory studies yield data of interest to field students of behavior or ecology, the potential for such contributions should not be underestimated. Indeed, field studies can contribute a wealth of unresolved questions that are most appropriately addressed by experiments in captivity. In addition, field studies may be essential to interpreting behavioral differences found among species or populations in captive studies of particular behavior patterns.

ACKNOWLEDGMENTS

This is contribution 83 from the graduate program in ecology and evolution at the State University of New York, Stony Brook. I thank Drs. Menzel, Wyers, and Gibbons for inviting me to speak on this topic and thus stimulating my appreciation of the importance of laboratory studies of behavior.

V. CONCLUSION

Naturalistic Environments in Captivity for Animal Behavior Research: Synthesis Statement

The contributions to this volume have addressed many of the administrative, regulatory, management, and research issues related to naturalistic environments in captivity. The reported research findings have demonstrated that naturalistic facilities can serve as a vital methological link between traditional animal laboratory environments and the wild, and can complement and add to empirical information obtained from these latter settings. Naturalistic habitats in captivity also offer a fundamental opportunity to understand how ecological systems interact with the biological and psychological processes of animals, and how to manage these systems in a way to ensure the physical health and psychological well-being of animals. It is clear that in the not too distant future there will be little distinction between the "wild" and managed naturalistic environments.

The task remains, however, to formulate what constitutes a naturalized research environment. It is also necessary to outline operational philosophies and methods that will enhance the development and maintenance of naturalistic facilities, to state the research and educational benefits of these environments, and to develop a protocol that will help in the resolution of problems that may arise as a result of conflict with existing federal regulatory standards. These questions were addressed on the last day of the 1987 conference. The following text is the product of discussions among the conference participants, and is their contribution to the present volume.

WHAT CONSTITUTES A NATURALIZED RESEARCH ENVIRONMENT?

Naturalized research environments functionally simulate or reproduce critical elements of the physical and social environment of a particular species, while affording investigators greater access to their subjects than is possible in the wild. Such environments are designed to promote the physical and psychological well-being of their inhabitants without compromising the safety of researchers or support staff. In contrast to standard laboratory environments, naturalistic facilities are characterized by one or more of the following features:

a. They usually exceed minimum USDA mandated space require-ments for a particular species .
b. They typically are outfitted with physical environmental fea-tures that are constructed from natural materials or are artistic reproductions in a variety of special configurations, such as vine tangles rather than rectangular PVC climbing structures.
c. Often in these environments the inhabitants are allowed to experience a variety of climatic and light conditions, although shelter from extremes is provided.
d. The inhabitants often have the opportunity to forage naturally or to obtain food beyond traditional provisioned diets.
e. Animals in naturalistic environments typically are maintained in social groups that allow for the expression of basic aspects of species normative behavior and social organization, including opportunities for reproduction and parenting. These social groups represent the unique multigenerational accumulation of genetic and cultural information that is not replicable.
f. Naturalistic environments may also contain more than one compatible species when required for management, research, or educational purposes.

DEVELOPMENT AND MAINTENANCE OF NATURALISTIC ANIMAL ENCLOSURES

The goal in developing and maintaining naturalistic animal facilities in captivity is to simulate the best available knowledge on the natural ecology of the animals while protecting their physical and psychological health. In order to achieve this goal it is imperative that the design of these facilities take into account the physiological and behavioral needs of the animals, the day-to-day management of the facility and sanitation of surfaces, and the scientific objectives of the staff. In particular, it is critical that there be proper air flow to preclude accumulation of toxins and particulates. There must be careful choice of substrates that minimize the accumulation of potential pathogens or hazardous conditions, yet permitting an acceptable simulated environment. Careful attention should also be devoted to the selection of building materials and the design of equipment to preclude offering habitats for unwanted arthropods and vermin. The suppression of pests can be best achieved by using strategies that take into account the behavior and ecology of pests in the natural environment.

The obligations of researchers extend beyond that of traditional science. Investigators must remain cognizant of the ways in which scientific endeavors may advance the disciplines of facility design and management. It is important that researchers, during the information phase of facility design, obtain detailed knowledge on the species' health requirements, behavior, and ecology. Investigators must also establish clear lines of authority and responsibility that will address issues regarding the scientific and management objectives of naturalistic facilities. It is also the responsibility of the scientific staff to ensure that occupational health standards are met, and to consider the "cost/benefit" institutional priorities in the operation of naturalistic facilities.

BENEFITS ASSOCIATED WITH THE USE OF NATURALISTIC FACILITIES: ANIMALS, RESEARCH, AND EDUCATION

Animals

The use of naturalistic animal facilities will encourage the development of species-specific criteria for the recognition of psychological well-being, and by doing so will promote the health and well-being of animals by providing natural individual and social groupings, and by reducing the need to handle animals for the management of health. These facilities will also promote the development of improved management and health care techniques, and the empirical scientific study of habitat design. Naturalistic facilities will also lead to the improvement of species propagation in captivity, and the development of social skills that are needed for the reintroduction of endangered species to suitable habitats in the wild.

Research

The research benefits of naturalistic animal environments in captivity are many and involve two general areas: research methodology, and knowledge regarding aspects of the animals' biological, psychological, and social functioning. It is also important to note that knowledge gained from studies in naturalistic facilities can be used to promote the conservation of species and the restoration of critical habitats in the wild.

Research Methodology. Naturalistic environments increase the reproducibility of studies because they take into account the biological and psychological needs of animals, and promote the occurrence of animals' behaviors in contexts that simulate environmental and social conditions in the wild. Further, naturalistic animal environments permit the testing of generality of phenomena observed in small cages to more natural conditions, including the wild. In naturalistic environments it is possible to study phenomena (such as psychological aspects of population density) that cannot be studied in traditional laboratory settings or in the wild. There will also be increased opportunity to conduct interdisciplinary studies of the same animal(s) by animal behaviorists, anthropologists, ecologists, physiologists, philosophers, and artists. Finally, researchers must include, in their scientific publications, more detailed information regarding the ecological and physical environmental features of the environment, and the management procedures used to maintain the facility. It is also important that publications document important failures in facility design and management as well as successes.

Knowledge of Biological and Psychological Functioning. The use of naturalistic environments in captivity in animal behavior research will promote the occurrence of natural and individual social behaviors from animals with known genealogies. Such investigations will also permit the detailing of animals' developmental and life history events, and the physiological, behavioral, and cognitive responses to such events. Studies involving animals' reproductive cycles and parental behaviors are two examples where such life history information would prove valuable.

Education

Naturalistic animal environments in captivity will allow for the training of students at all educational levels in observational methodologies. These environments will also foster an appreciation for animals difficult to achieve in traditional laboratory settings. They can also be used to promote the need to preserve species in the wild and their ecosystems. There is hardly an academic subject that, in some way, cannot use naturalistic animal facilities as an educational tool.

RESOLUTION OF PROBLEMS WITH REGULATORY STANDARDS

Because of the variability in design features that may exist between naturalistic environments, it is important that administrative and scientific personnel at each institution define exactly what is meant by a naturalistic environment (e.g., indoor, indoor/outdoor, outdoor with shelter, outdoor). It is also important for scientists to communicate the importance of their research and scientific needs to department heads, administrators, and the public. Institutions must also take an organized approach when recruiting faculty, and to incorporate their research needs in existing and future administrative, economic, and operational plans. Veterinarians must meet with new investigators and work through their respective requirements. Institutions should also develop an animal care and use program that incorporates the facility into the academic plan, and assists administrators in the financial support of research in these facilities, and publicly backs researchers when necessary. Finally, it is important that administrators and researchers work with funding agencies prior to the submission of grants when "exceptional" facilities are involved.

Other measures that may be taken to minimize the occurrence of problems associated with naturalistic environments involve the development of a brochure to guide faculty and administrators to the management and scientific goals and philosophies underlying the operation of the facility. It is also recommended that a proactive approach be taken to the review and approval of the facility by regulatory personnel. This process, however, must be defensible. Encouragement should also be given to researchers to work with national groups interested in naturalistic observation (e.g., the American Psychological Association, American Society of Primatologists, and Animal Behavior Society).

In conclusion, institutional administrators and researchers should petition that there be immediately added to USDA regulations the fact that unconventional facilities be allowed. There should also be a follow-up to the petition by educating USDA inspectors how "nonstandard" design features can enhance the health and well-being of animals. It is also recommended that regulatory personnel be educated as to the need of acceptable guidelines for the inclusion of substrates such as branches and bedding, and for the extermination of pests such as cockroaches.

References

Adams, N., and R. Boice. 1981. Mouse (*Mus*) burrows: effects of age, strain, and domestication. *Animal Learning and Behavior* 9:140–144.

Adams, N., and R. Boice. 1983. A longitudinal study of dominance in an outdoor colony of domestic rats. *Journal of Comparative Psychology* 97:24–33.

Ader, R. 1967. The influence of psychological factors on disease susceptibility in animals. Pages 219–238 in M.L. Conalty, ed., *Husbandry of laboratory animals*. Academic Press, London.

Adolph, E.F. 1956. General and specific characteristics of physiological adaptations. *American Journal of Physiology* 184:18–28.

Ågren, G. 1984. Pair formation in the Mongolian gerbil. *Animal Behaviour* 32:528–535.

Ågren, G., Q. Zhou, and W. Zhong. 1989. Ecology and social behaviour of Mongolian gerbils, *Meriones unquiculatus*, at Xilinhoit, Inner Mongolia, China. *Animal Behaviour* 37:11–27.

Alexander, B.K., and E.M. Roth. 1971. The effects of acute crowding on aggressive behavior of Japanese monkeys. *Behaviour* 39:73–89.

Altmann, J., and P. Muruthi. 1988. Differences in daily life between semiprovisioned and wild–feeding baboons. *American Journal of Primatology* 15:213–221.

Altman, J. 1974. Observational study of behavior: sampling methods. *Behaviour* 49:227–267.

Altmann, S.A. 1965. Sociobiology of rhesus monkeys. II. Stochastics of social communication. *Journal of Theoretical Biology* 8:490–522.

Altmann, S.A., and J. Altmann. 1979. Demographic constraints on behavior and social organization. Pages 47–63 in I.S. Bernstein and E.O. Smith, eds., *Primate ecology and human origins: Ecological influences on social organization*. Garland Press, New York.

Amlander, C.J., and D.W. MacDonald, eds. 1980. *A handbook on biotelemetry and radio tracking.* Pergamon Press, Oxford.

Andelman, S.J. 1985. Ecology and reproductive strategies of vervet monkeys (*Cercopithecus aethiops*) in Amboseli National Park, Kenya. Ph.D. Dissertation. University of Washington (*Dissertation Abstracts International* 47(7)B:2229).

Anderson, K.V., F.P. Coyle, and W.K. O'Steen. 1972. Retinal degeneration produced by low-intensity colored light. *Experimental Neurology* 35:233–238.

Andrews, M.W., and Rosenblum, L.A. 1988. Relationship between foraging and affiliative social referencing in primates. Pages 247–268 in J.E. Fa and C.H. Southwick, eds., *Ecology and behavior of food-enhanced primate groups.* Alan R. Liss, New York.

Anisko, J.J., S.F. Suer, M.K. McClintock, and N.T. Adler. 1978. Relation between 22-kHZ ultrasonic signals and sociosexual behavior in rats. *Journal of Comparative and physiological Psychology* 92:821–829.

Anisman, H., and T.G. Waller. 1973. Effects of inescapable shock on subsequent avoidance performance: role of response repertoire changes. *Behavioral Biology* 9:331.

Anonymous. 1966. *Animal welfare act* 7:2131–2157, United States Congress, Washington, D.C.

Anonymous. 1985. *Animal welfare act regulations, 9 CFR Chapter 1, Subchapter A-animal Welfare.* APHIS, U.S. Department of Agriculture, Bethesda, Maryland.

Anonymous. 1986. *Public Health Service Policy on Humane Care and Use of laboratory animals.* Office of Protection from research risks. Department of Health and Human Services, Public Health Service, and National Institutes of Health, Bethesda, Maryland.

Anonymous. 1988. *Guide for the care and use of agricultural animals in agricultural research and teaching,* 1st. ed. Consortium for Developing a Guide for the Care and Use of Agricultural Animals in Agricultural Research and Teaching. National Association of State Universities and Land-Grant Colleges, Washington, D.C.

Anonymous. 1989. *American society for heating, refrigerating and air conditioning engineers handbook and product directory. Fundamentals Volume,* pages 9.1–9.18. American Society for Heating, Refrigerating and Air Conditioning Engineers, Atlanta, Georgia.

Anonymous. 1990. *Animal welfare enforcement for fiscal year 1989.* U.S. Department of Agriculture, APHIS-REAC, Bethesda, Maryland.

Appleman, R.D., and J.C. Delouche. 1958. Behavioral, physiological and biochemical responses of goats to temperature, 0° to 4°C. *Journal of Animal Science* 17:326–335.

Arling, G.L., and H.F. Harlow. 1967. Effects of social deprivation on maternal behavior of rhesus monkeys. *Journal of Comparative and Physiological Psychology* 64:371–377.

Arnone, M., and R. Dantzer. 1980. Does frustration induce aggression in pigs? *Applied Animal Ethology* 6:351–362.

Atwell, J. K. 1987. Veterinary services in emergencies: epizootic and zoonotic diseases. *Journal of the American Veterinary Medical Association* 190:709–713 .

Ayres, F.A., and S.J. Arnold. 1983. Behavioral variability in natural populations. IV. Mendelian models and heritability of a feeding response in the garter snake, *Thamnophis elegans*. *Heredity* 51:405–413.

Bader, R.S. 1956. Variability in wild and inbred mammalian populations. *Quarterly Journal of the Florida Academy of Science* 19:14–34.

Baetjer, A.M. 1968. Role of environmental temperature and humidity in susceptibility to disease. *Archives of Environmental Health* 16:565–570.

Baker, A., J. Dietz, and D. Kleiman. In press. Behavioral evidence for monogamy in multi-male groups of golden lion tamarins. *Animal Behaviour*.

Baldo, B.A., and R.C. Panzani. 1988. Detection of IgE antibodies to a wide range of insect species in subjects with suspected inhalant allergies to insects. *International Archives of Allergy and Applied Immunology* 85:278–287.

Balk, M.W. 1983. Overview of the state of the art in health monitoring. Pages 3–16 in E.C. Melby, Jr. and M.W. Balk, eds., *The importance of laboratory animal genetics, health, and the environment in biomedical research*. Academic Press, Orlando, Florida.

Ballard, J.B., and R.E. Gold. 1982. The effect of selected baits on the efficacy of a sticky trap in the evaluation of German cockroach populations. *Journal of the Kansas Entomological Society* 55:86–90.

Bancroft, L. S. 1985. Computerisation of the animal house. *Animal Technology* 36:191–138.

Banks, E.M. 1982. Behavioral research to answer questions about animal welfare. *Journal of Animal Science* 54:434–446 .

Bannikov, A. G. 1954 . The places inhabited and natural history of *Meriones unguiculatus*. Pages 410–415 in *Mammals of the Mongolian Peoples Republic*. U.S.S.R. Academy of Sciences.

Baran, D., and S. E. Glickman. 1967. Some determinants of shredding activity in the Mongolian gerbil. Paper presented at the annual meetings of the Eastern Psychological Association, Washington, D.C.

Baran, D., and S. E. Glickman. 1970 . "Territorial marking" in the Mongolian gerbil: a study of sensory control and function. *Journal of Comparative and Physiological Psychology* 71: 237–245.

Bard, R.L., and L. Kurlantzick. 1990. Further commentary on Fox's *In the Matter of "Baby M."* *Politics and the Life Sciences* 9:130–134.

Barnett, S.A. 1958. Physiological effects of "social stress" in wild rats—I. The adrenal cortex. *Journal of Psychosomatic Research* 3:1–11.

Barrett, A.M., and M.A. Stockham. 1965. The response of the pituitary-adrenal system to a stressful stimulus: the effect of conditioning and pentobarbitone treatment. *Journal of Endocrinology* 33:145–152.

Bateson, P. 1981. Behavioral development and the evolutionary process. Pages 133–151 in K. Immelman, G.W. Barlow, L. Petranovich, and M. Main, eds. *Behavior and development*. Cambridge University Press, Cambridge, England.

Bauers, K.A., and C.T. Snowdon. 1990. Discrimination of chirp vocalizations in the cotton-top tamarin. *American Journal of Primatology* 21:53–60.

Baum, W.M. 1983. Studying foraging in the psychological laboratory. Pages 253–283 in R.L. Mellgren, ed., *Animal cognition and behavior*. North-Holland, Amsterdam.

Bayne, K., S. Dexter, H. Mainzer, C. McCully, G. Campbell, and F. Yamada. 1992. The use of artificial turf as a foraging substrate for individually housed rhesus monkeys (*Macaca mulatta*). *Animal Welfare* 1:39–53.

Bayne, K., H. Mainzer, S. Dexter, G. Campbell, F. Yamada, and S. Suomi. 1991. The reduction of abnormal behaviors in individually housed rhesus monkeys (*Macaca mulatta*) with a foraging/grooming board. *American Journal of Primatology* 23:23–36.

Beauchamp, G.K., K. Yamazaki, and E.A. Boyse. 1985. The chemosensory recognition of genetic individuality. *Scientific American* 253:86–92.

Beck, B. 1980. *Animal Tool Behavior*. New York, Garland STPM Press.

Beck, B., D. Kleiman, I. Castro, B. Rettberg-Beck, and C. Carvalho. In press. Preparation of captive-born golden lion tamarins for release into the wild. In D. Kleiman, ed., *A case study in conservation biology: The golden lion tamarin.* Smithsonian Institution Press, Washington, D.C.

Beck, B., and M.L. Power. 1988. Correlates of sexual and maternal competence in captive gorillas. *Zoo Biology* 7: 339–350.

Beckley, S., and M. Novak. 1989. An examination of various foraging components and their suitability as enrichment tools for captively housed primates. *American Journal of Primatology, Supplement* 1:37–43.

Behbehani, A.M. 1972. *Human, viral, bedsonial and rickettsial diseases.* Charles C. Thomas, Springfield, Illinois.

Bellhorn, R.W. 1980. Lighting in the animal environment. *Laboratory Animal Science* 30:440–450 .

Belzung, C., and J.R. Anderson. 1986. Social rank and responses to feeding competition in rhesus monkeys. *Behavioral Processes* 12:307–316.

Benedict, F.G. 1938. *Vital energetics: A study in comparative basal metabolism.* Publication no. 503. Carnegie Institute of Washington, Washington, D.C.

Benirschke, K., B. Lasley, and O. Ryder. 1980. The technology of captive propagation. Pages 225–242 in M. E. Soulé and B.A. Wilcox, eds., *Conservation biology: An evolutionary-ecological perspective.* Sinauer Associates, Sunderland, Massachusetts.

Bennett, G.W., and J.M. Owens, eds. 1986. *Advances in urban pest management.* Van Nostrand Reinhold, New York.

Bennett, G.W., J.W. Yonker, and E. S. Runstrom. 1986. Influence of hyproprene on German cockroach (Dictyoptera: Blattellidae) populations in public housing. *Journal of Economic Entomology* 79:1032–1035.

Bercovitch, F.B., and T.E. Ziegler. 1989. Reproductive strategies and primate conservation. *Zoo Biology, Supplement* 1: 163–169.

Berg, J.K. 1987. Behavior of the Japanese serow (*Capricornis crispus*) at the San Diego Wild Animal Park. Pages 165–181 in H. Soma, ed., *The biology and management of Caricornis and related mountain antelopes.* Croom Helm, London.

Berkson, G. 1967. Abnormal stereotyped motor acts. Pages 76–94 in J. Zubin and H.F. Hunt, eds., *Comparative Psychopathology: Animal and human.* Grune & Stratton, New York.

Berkum, F.H. van, and J.S. Tsuji. 1987. Inter-familiar differences in sprint speed in hatchling *Sceloporus occidentalis* (Reptilia: Iguanidae). *Journal of Zoology: Proceedings of the Zoological Society of London* 212 :511–519 .

Bernstein, I.S., T.P. Gordon, and R.M. Rose. 1974. Factors influencing the expression of aggression during introductions to rhesus monkey groups. Pages 211–240 in R.L. Holloway, ed., *Primate aggression, territoriality, and xenophobia.* Academic Press, New York.

Berry, R.J. 1969. The genetical implications of domestication in animals. Pages 207–217 in P.J. Ucko and G.W. Dimbleby, eds., *The domestication and exploitation of plants and animals.* Aldine, Chicago.

Besch, E.L. 1975. Animal cage room dry-bulb and dew-point temperature differentials. *American Society for Heating, Refrigerating and Air Conditioning Engineers Transactions* 81: 549–558 .

———. 1980. Environmental quality within animal facilities. *Laboratory Animal Science* 30:385–406 .

———. 1985. Definition of laboratory animal environmental conditions. Pages 297–315 in G.P. Moberg, ed., *Animal stress.* American Physiological Society, Bethesda, Maryland.

———. 1990. Environmental variables and animal needs. Pages 113–131 in B.L. Rollin, ed., *The experimental animal in biomedical research,* volume 1. CRC Press, Boca Raton, Florida.

Bielitzki, J., T.G. Susor, K. Elias, and D.M. Dowden. 1990. Improved cage design for single housing of social nonhuman primates. *Laboratory Animal Science* 40:428–431.

Birdsall, D.A., and D. Nash. 1973. Occurrence of successful multiple insemination of females in natural populations of deer mice (*Peromyscus maniculatus*). *Evolution* 27:106 –110.

Blalock, J. E. 1984. The immune system as a sensory organ. *Journal of Immunology* 132:1067–1070.

Blanchard, D.C., and R.J. Blanchard. 1990. The colony model of aggression and defense. Pages 410–430 in D.A. Dewsbury, ed., *Contemporary issues in comparative psychology.* Sinauer Associates, Sunderland, Massachusetts.

Blanchard, R.J., P.F. Brain, P. C. Blanchard, and S. Parmigiani, eds. 1989. *Ethoexperimental approaches to the study of behavior.* Kluwer, Dordrecht.

Blanchard, R.J., K. J. Flannelly, and D.C. Blanchard. 1988. Life-span studies of dominance and aggression in established colonies of laboratory rats. *Physiology & Behavior* 43:1–7.

Blanchard, D.C., C. Fukunaga-Stinson, L.K. Takahashi, K.J. Flannelly, and R.J. Blanchard. 1984. Dominance and aggression in social groups of male and female rats. *Behavioural Processes* 9:31–48.

Bloomsmith, M.A., P.L. Alford, and T.L. Maple. 1988. Successful feeding enrichment for captive chimpanzees. *American Journal of Primatology* 16:155–164.

Bloomstrand, M., K. Riddle, P. Alford, and T.L. Maple. 1986. Objective evaluation of an enrichment device for captive chimpanzees (*Pan troglodytes*). *Zoo Biology* 5:293–300.

Boccia, M.L., M. Laudenslager, and M. Reite. 1988. Food distribution, dominance, and aggressive behaviors in bonnet macaques. *American Journal of Primatology* 16:123–130.

Boever, W.J. 1972. Mismanagement trauma. *Journal of Zoo Animal Medicine* 3:20–25.

Boice, R. 1973. Domestication. *Psychological Bulletin* 80:215–230.

———. 1977. Burrows of wild and albino rats: effects of domestication, outdoor raising, age, experience, and maternal state. *Journal of Comparative and Physiological Psychology* 91:649–661

———. 1981. Captivity and feralization. *Psychological Bulletin* 89:407–421.

Boice, R., and N. Adams. 1980. Outdoor enclosures for feralizing rats and mice. *Behavior Research Methods & Instrumentation* 12:577–582.

Boinski, S., and D.M. Fragaszy. 1989. The ontogeny of foraging in squirrel monkeys, *Saimiri oerstedi*. *Animal Behaviour* 37:415–428.

Bolles, R.C. 1989. Acquired behaviors: aversive learning. Pages 167–179 in R.J. Blanchard, P.F. Brain, D.C. Blanchard and S. Parmigiani, eds., *Ethoexperimental approaches to the study of behavior*. Kluwer, Dordrecht.

Borror, D.J., D.M. Delong, and C.A. Triplehorn. 1976. *An introduction to the study of insects*, 4th. ed. Holt, Rinehart and Winston, New York.

Bowen, W.D. 1978. Social organization of the coyote in relation to prey size. Ph.D. Dissertation. University of British Columbia (*Disseration Abstracts International* 40(4)B:1500).

Bowen, W.D., and I. McTaggart Cowan. 1980. Scent marking in coyotes. *Canadian Journal of Zoology* 58:473–480.

Bowlby, J. 1951. Maternal care and mental health. *Bulletin of the World Health Organization* 3:355–534.

Bowman, J.C. 1977. *Animals for man.* Edward Arnold, London.

Bradley, J.V. 1968. *Distribution-free statistical tests.* Prentice-Hall, Englewood Cliffs, New Jersey.

Bracewell, R.J., and A.H. Black. 1974. The effects of constraint and non-contingent preshock on subsequent escape learning in the rat. *Learning and Motivation* 5:53–69.

Brain, P.F. 1983. Pituitary-gonadal influences on social aggression. Pages 3–25 in B.B. Svare, ed., *Hormones and aggressive behavior.* Plenum Press, New York.

Brambell, M.R. 1977. Reintroduction. *International Zoo Yearbook* 17: 112–116.

Brant, D.H., and J.L. Kavanau. 1964. 'Unrewarded' exploration and learning of complex mazes by wild and domestic mice. *Nature* 204:267–269.

Brant, D.H., and J.L. Kavanau. 1965. Exploration and movement patterns of the canyon mouse *Peromyscus crinitus* in an extensive laboratory enclosure. *Ecology* 46:452–461.

Brenner, R.J. 1988. Focality and mobility of some peridomestic cockroaches in Florida (Dictyoptera: Blattaria). *Annals of the Entomological Society of America* 81:581–592.

―――. 1991. Asian cockroaches: implications to the food industry and complexities of management strategies. Pages 121–130 in J.R. Gorham, ed., *Ecology and management of food-industry pests.* FDA Technical Bulletin 4, Association of Official Analytical Chemists, Arlington, Virginia.

Brenner, R.J., P.G. Koehler, and R.S. Patterson. 1987. Implications of cockroach infestations to human health. *Infections in Medicine* 4:349–355, 358–359, 393.

Brenner, R.J., P.G. Koehler, and R.S. Patterson. 1988. Integration of fenoxycarb into a German cockroach (Orthoptera: Blattellidae) management program. *Journal of Economic Entomology* 81:1404–1407.

Brenner, R.J., and R.S. Patterson. 1988. Efficiency of a new trapping and marking technique for peridomestic cockroaches (Dictyoptera: Blattaria). *Journal of Medical Entomology* 25:489–492.

Brenner, R.J., and R.S. Patterson. 1989. Laboratory feeding activity and bait preferences of four species of cockroaches (Orthoptera: Blattaria). *Journal of Economic Entomology* 82:159–162.

Brenner, R.J., R.S. Patterson, and P.G. Koehler. 1988. Ecology, behavior, and distribution of *Blattella asahinai* (Orthoptera: Blattellidae) in central Florida. *Annals of the Entomological Society of America* 81:432–436.

Broderson, J.R., J.R. Lindsey, and J.E. Crawford. 1976. The role of environmental ammonia in respiratory mycoplasmosis of rats. *American Journal of Pathology* 85:115–130.

Brody, 5. 1974. *Bioenergetics and growth*. Reinhold, New York.

Broner, N. 1978. Low frequency and infrasonic noise in transportation. *Applied Acoustics* 11:129–146.

Bronikowski, E., Beck, B., and M. Power. 1989. Innovation, exhibition and conservation: free-ranging tamarins at the National Zoological Park. *Proceedings of the Annual Meeting of the American Associa tion of Zoological Parks and Aquariums*: 540–546.

Bronson, F.H. 1979. The reproductive ecology of the house mouse. *Quarterly Review of Biology* 54:265–299.

Broom, D.M. 1987. The veterinary relevance of farm animal ethology. *Veterinary Record* 121:400–402.

Brown, C.W., and E.E. Ghiselli. 1950. *Scientific Method in Psychology*. McGraw-Hill, New York.

Brown, J.H., and R.C. Lasiewski. 1972. Metabolism of weasels: the cost of being long and thin. *Ecology* 53:939–943.

Bruce, H.M. 1959. An exteroceptive block to pregnancy in the mouse. *Nature* 184:105.

Brunswik, E. 1951. Organismic achievement and environmental probability. Pages 188–202 in M. H. Marx, ed., *Psychological theory: Contemporary readings*. Macmillan, New York.

———. 1955a. Representative design and probabilistic theory in a functional psychology. *Psychological Review* 62:193–217.

———. 1955b. In defense of probabilistic functionalism: a reply. *Psychological Review* 62:236–242.

Buck, R., and B. Ginsburg. 1991. Spontaneous communication and altruism: the communicative gene hypothesis. Pages 149–175 in M.S.

Clark, ed., *Review of personality and social psychology, Volume 12.* Sage Publications, Newbury Park, California.

Bundy, D.A., and J.H. Steele. 1984. Parasitic zoonoses in the Caribbean Region: a review. *International Journal of Zoonosis* 11:1–38.

Bünning, E. 1967. *The physiological clock: Circadian rhythms and biological chronometry,* 2nd. ed., pages 126–129. Springer-Verlag, New York.

Burek, J.D., and B.A. Schwetz. 1980. Considerations in the selection and use of chemicals within the animal facility. *Laboratory Animal Science* 30:414–421.

Bush, M., L.G. Phillips, and R.J. Montali. 1987. Clinical management of captive tigers. Pages 171–204 in R.L. Tilson and U. S . Seal, eds., *Tigers of the world. The biology, biopolitics, management, and conservation of an endangered species.* Noyes Publications, Park Ridge, New Jersey.

Byrne, G.D., and S.J. Suomi. 1991. Effects of woodchips and buried food on behavior patterns and psychological well-being of captive rhesus monkeys. *American Journal of Primatology* 23:141–152.

Caldwell, G.S., S.E. Glickman, and E.R. Smith. 1984. Seasonal aggression independent of seasonal testosterone in wood rats. *Proceedings of the National Academy of Sciences, U.S.A.* 81:5255–5257.

Calhoun, J. B. 1962a. *The ecology and sociology of the Norway rat.* Public Health Service, Bethesda, Maryland.

———. 1962b. A "behavioral sink." Pages 295–315 in E.L. Bliss, ed., *Roots of behavior: Genetics, instinct, and socialization in animal behavior.* Harper & Brothers, New York.

———. 1963. Population density and social pathology. *Scientific American* 206:139–148.

Canguilhem, B. 1987. Neural and neuroendocrine basis of mammalian hibernation and heterothermy. Pages 136–151 in P. Pevet, ed., *Comparative physiology of environmental adaptations,* volume 3. Karger, Basel, Switzerland.

Carlstead, K., J. Seidensticker, and R. Baldwin. 1991. Environmental enrichment for zoo bears. *Zoo Biology* 10:3–16.

Carnahan, R.B.A. 1958. Keratoconjunctivitis in broiler chicks. *Veterinary Record* 70:35–37.

Carter, J. 1988. Survival training in chimps. *Smithsonian* 19:36–39.

Carter, C.S., L.L. Getz, and M. Cohen-Parsons. 1986. Relationships between social organization and behavioral endocrinology in a monogamous mammal. *Advances in the Study of Behavior* 16:109–145.

Chamove, A.S. 1984. Role of vision in social development in monkeys. *Child Development* 55:1394–1411.

Chamove, A.S., J.R. Anderson, S.C. Morgan-Jones, and S.P. Jones. 1982. Deep woodchip litter: hygiene, feeding and behavioral enhancement in eight primate species. *International Journal for the Study of Animal Problems* 3:308–318.

Champion, F.W. 1934. *The jungle in sunlight and shadow*. Charles Scribner's, New York.

Chapel, R.N., and R.J. Hudson. 1978. Winter bioenergetics of Rocky Mountain bighorn sheep. *Canadian Journal of Zoology* 56:2388–2393.

Chapman, C. 1988. Patch use and patch depletion by the spider and howling monkeys of Santa Rosa National Park, Costa Rica. *Behaviour* 105:99–116.

Chapman, F.M. 1912. Handbook of birds of eastern North America. D. Appleton, New York.

———. 1929. *My tropical air castle*. D. Appleton-Century, New York.

Chapman, R.F. 1971. *The insects: Structure and function*, 2nd. ed. Elsevier, North Holland, New York.

Cheesman, C.L., and R.B. Mitson, eds. 1982. *Telemetric studies of vertebrates. Symposia of the Zoological Society of London, Number 49.* Academic Press, London.

Cheney, D.L., and R.M. Seyfarth. 1981. Selective forces affecting the predator alarm calls of vervet monkeys. *Behaviour* 76:25–61.

Cheney, D.L., and R.M. Seyfarth. 1985. Vervet monkey alarm calls: manipulation through shared information? *Behaviour* 94:150–166.

Chepko-Sade, D.B., and D.S. Sade. 1979. Patterns of group splitting within matrilineal kinship groups. A study of social group structure in *Macaca mulatta*. *Behavioral Ecology and Sociobiology* 5:67–86.

Cherry, J.A., M.K. Izard, and E.L. Simons. 1987. Description of ultrasonic vocalizations of the mouse lemur (*Microcebus murinus*) and the fat-tailed dwarf lemur (*Cheirogaleus medius*). *American Journal of Primatology* 13:181–185.

Christian, J.J. 1955. Effect of population size on the adrenal glands and reproductive organs of male mice in populations of fixed size. *American Journal of Physiology* 182:292–300.

Christian, J.J., and C.D. Lemunyan. 1958. Adverse effects of crowding on lactation and reproduction of mice and two generations of their progeny. *Endocrinology* 63:517–529.

Cinti, D.L., M.A. Lemelin, and J. Christian. 1976. Induction of liver microsomal mixed-function oxidases by volatile hydrocardons. *Biochemical Pharmacology* 25:100–103.

Clark, A.B. 1978a. Olfactory communication, *Galago crassicaudatus* and the social life of prosimians. Pages 109–117 in D.J. Chivers and J. Hergert, eds., *Recent advances in primatology, II. Evolution.* Academic Press, New York.

———. 1978b. Sex ratio and local resource competition in a prosimian primate. *Science* 201:163–165.

Clarke, A.M., and A.D.B. Clarke, eds. 1976. *Early experience: Myth and evidence.* Free Press, New York.

Clark, M.M., and B.G. Galef, Jr. 1977. The role of the physical rearing environment in the domestication of the Mongolian gerbil (*Meriones unguiculatus*). *Animal Behaviour* 25:298–316.

Clark, M.M., and B.G. Galef, Jr. 1980. Effects of rearing environment on adrenal weights, sexual development, and behavior in gerbils: an examination of Richter's domestication hypothesis. *Journal of Comparative and Physiological Psychology* 94:857–863.

Cleveland, J., and C.T. Snowdon. 1982. The complex vocal repertoire of the adult cotton-top tamarin (*Saguinus oedipus oedipus*). *Zeitschrift für Tierpsychologie* 58:231–270.

Cleveland, J., and C.T. Snowdon. 1984. Social development during the first twenty weeks in the cotton-top tamarin (*Saguinus o. oedipus*). *Animal Behaviour* 32:432–444.

Clough, G. 1982. Environmental effects on animals used in biomedical research. *Biological Reviews* 57:487–523.

Clough G., and M.R. Gamble. 1976. *Laboratory animal houses: A guide to the design and planning of animal facilities.* Laboratory Animal Care manual series no. 4, Environmental Physiology Department,

Carshalton, England: Medical Research Council, Laboratory Animals Centre.

Clulow, F.V., E.A. Franchetto, and P.E. Langford. 1982. Pregnancy failure in the red-backed vole, *Clethrionomys gapperi*. *Journal of Mammology* 63:499–500.

Clutton-Brock, T.H. 1982. Sons and daughters. *Nature* 298:11–13.

Clutton-Brock, T.H., and P.H. Harvey. 1977. Species differences in feeding and ranging behavior in primates. Pages 557-584 in T.H. Clutton-Brock, ed., *Primate ecology: Studies of feeding and ranging behaviour in lemurs, monkeys and apes*. Academic Press, London.

Cochran, D.G. 1975. Excretion in insects. Pages 177–281 in D.J. Candy and B.A. Kilby, eds., *Insect biochemistry and function*. Chapman and Hall, London.

————. 1989. Monitoring for insecticide resistance in field-collected strains of the German cockroach (Dictyoptera: Blattellidae). *Journal of Economic Entomology* 82:336–341.

Cochran, D.G., and D.E. Mullins. 1982. Physiological processes related to nitrogen excretion in cockroaches. *Journal of Experimental Zoology* 222:277–285.

Code of Federal Regulations, CFR. 1982. Laboratory Animal Welfare Act, 1966 (Public Law 89–544). Amended 1970 (Public Law 91–579), Amended 1976 (Public Law 94–279). Code of Federal Regulations, Title 9, Animals and Animal Products, Subchapter A—Animal Welfare. *Federal Register* 54(168), 31 August 1989. U.S. Government Printing Office, Washington, D.C.

Coe, C.L., L.T. Rosenberg, M. Fischer, and S. Levine. 1987. Psychological factors capable of preventing the inhibition of antibody responses in separated infant monkeys. *Child Development* 58:1420–1430.

Coe, C.L., and J. Scheffler. 1989. Utility of immune measures for evaluating psychological well-being in nonhuman primates. *Zoo Biology, Supplement* 1:89–99.

Coe, J.C. 1989. Naturalizing habitats for captive primates. *Zoo Biology, Supplement* 1:117–125.

Cohen, J. 1977. *Statistical power analysis for the behavioral sciences*. Academic Press, New York.

Colam-Ainsworth, P., G.A. Lunn, R.C. Thomas, and R.G. Eddy. 1989. Behavior of cows in cubicles and its possible relationship with laminitis in replacement dairy heifers. *Veterinary Record* 125:573–575.

Colgan, P.W. ed. 1978. *Quantitative Ethology*. John Wiley Sons New York.

Collier, G.H. 1982. Determinants of choice. *Nebraska Symposium on Motivation* 29:69–127.

Collier, G., and C. Rovee-Collier. 1981. A comparative analysis of optimal foraging behavior: laboratory simulations. Pages 39–76 in A. Kamil and T. Sargent, eds., *Foraging behavior: Ecological, ethological, and psychological approaches.* Garland STPM Press, New York.

Collier, G.H., and C.K. Rovee-Collier. 1983. An ecological perspective of reinforcement and motivation. Pages 427–441 in E. Satinoff and P. Teitelbaum, eds. *Handbook of behavioral neurobiology: Motivation.* Plenum Press, New York.

Colvin, J.D. 1986. Proximate causes of male emigration at puberty in rhesus monkeys. Pages 131–157 in R.G. Rawlins and M.J. Kessler, eds., *The Cayo Santiago macaques: History, behavior & biology.* State University of New York Press, Albany, New York.

Conney, A.H., and J.J. Burns. 1972. Metabolic interactions among environmental chemicals and drugs. *Science* 178:576–586.

Cook, T.D., and D.T. Campbell. 1979. *Quasi-experimentation: Design and analysis for field settings.* Rand McNally, Chicago.

Cooper, E.C. 1989. Design considerations for research animal facilities. *Lab Animal* 18:23–26.

Cornwell, P.B. 1968. *The cockroach*, volume 1. Hutchinson, London.

Cranefield, P.F. 1974. *The way in and the way out: Francois Magendie, Charles Bell and the roots of the spinal nerves.* Future, Mount Kisco, New York.

Crowcroft, P. 1966. *Mice all over.* Dufour, Chester Springs, Pennsylvania.

Croxton, F.E., and D.J. Cowden. 1946. *Applied general statistics.* Prentice-Hall, New York.

Cummings, J.F. 1986. An employee appraisal model for supervisors and managers. *Lab Animal* 15:19–23.

Cunliffe-Beamer, T.L., L.C. Freeman, and D.D. Meyers. 1981. Barbiturate sleeptime in mice exposed to autoclaved or unautoclaved wood beddings. *Laboratory Animal Science* 31:672--675.

Cunningham, C.L., and J. Peris. 1983. A microcomputer system for temperature biotelemetry. *Behavior, Research Methods, and Instrumentation* 15:598–603.

Curtis, S.E. 1985. What consititutes animal well-being? Pages 1–14 in G.P. Moberg, ed., *Animal Stress*. American Physiological Society, Bethesda, Maryland.

Dantzer, R. 1986. Behavioral, physiological and functional aspects of stereotyped behavior: a review and a reinterpretation. *Journal of Animal Science* 62:1776–1786.

Dantzer, R., and P. Mormede. 1983. Stress in farm animals: a need for reevaluation. *Journal of Animal Science* 57:6–18.

Darling, F.F. 1937. *A herd of red deer*. Oxford University Press, London.

Darwin, C. 1872. *The expression of the emotions in man and animals*. University of Chicago Press (reprinted 1965), Chicago.

Dashbach, N.J., M.W. Schein, and D.E. Haines. 1982. Cage-size effects on locomotor, grooming, and agonistic behaviors of the slow loris, *Nycticebus coucang* (Primates, Lorisidae). *Applied Animal Ethology* 9:317–330.

Dauterman, W.C. 1980. Physiological factors affecting metabolism of xenobiotics. Pages 133–142 in E. Hodgson and F. E. Guthrie, eds., *Introduction to Biochemical Toxicology*. Elsevier, New York.

Davenport, J. 1985. *Environmental stress and behavioural adaptation*. Croom Helm, London.

Davey, F.K. 1965. Hair humidity elements. Pages 571–573 in R.E. Ruskin, ed., *Humidity and moisture: Measurement and control in science and industry*, volume one. *Principles and methods of measuring humidity in gases*. Reinhold, New York.

Davis, D.E. 1978. Social behavior in a laboratory environment. Pages 44–64 in E. Besch, Chairman, *Laboratory animal housing*. Institute of Laboratory Animal Resources, National Academy of Sciences, Washington, D.C.

Davis, D.E., and C.P. Read. 1958. Effect of behavior on development of resistance in trichinosis. *Proceedings of the Society for Experimental Biology and Medicine* 99:269–272.

Dawkins, M.S. 1980. *Animal suffering: The science of animal welfare.* Chapman and Hall, London.

———. 1983. Battery hens name their price: consumer demand theory and the measurement of ethological 'needs'. *Animal Behaviour* 31:1195–1205.

———. 1984. The many faces of animal suffering. Pages 298–303 in G. Ferry, ed., *The understanding of animals.* Basil Blackwell, Oxford.

Dawkins, R. 1976. *The selfish gene.* Oxford University Press, New York.

Denenberg, V.H., and E.M. Banks. 1969. Techniques of measurement and evaluation. Pages 192–233 in E.S.E. Hafez, ed., *The behaviour of domestic animals.* Williams and Wilkens, Baltimore, Maryland.

Dewsbury, D.A. 1979. Copulatory behavior of deer mice (*Peromyscus maniculatus*): II. A study of some factors regulating the find structure of behavior. Journal of Comparative and Physiological Psychology 93:161–177.

———. 1981a. Social dominance, copulatory behavior, and differential reproduction in deer mice (*Peromyscus maniculatus*). Journal of Comparative and Physiological Psychology 95:880–895.

———. 1981b. An excercise in the prediction of manogamy in the field from laboratory data on 42 species of muroid rodents. *The Biologist* 63:138–162.

———. 1982a. Dominance rank, copulatory behavior, and differential reproduction. *Quarterly Review of Biology* 57:135–139.

———. 1982b. Ejaculate cost and male choice. *American Naturalist* 119:601–610.

———. 1983. A comparative study of rodent social behavior in a seminatural enclosure. *Aggressive Behavior* 9:207–215.

———. 1984. Aggression, copulation, and differential reproduction of deer mice (*Peromyscus maniculatus*) in a seminatural enclosure. *Behaviour* 91:1–23.

———. 1987. Laboratory research on behavioral interactions as generators of population phenomena in rodents. *American Zoologist* 27:941–951.

———. 1988a. The comparative psychology of monogamy. *Nebraska Symposium on Motivation* 35:1–50.

————. 1988b. Kinship, familiarity, aggression and dominance in deer mice (*Peromyscus maniculatus*) in seminatural enclosures. *Journal of Comparative Psychology* 102:124–128.

————. Fathers and sons: genetic factors and social dominance in deer mice, *Peromyscus maniculatus. Animal Behaviour* 39:284–289.

————. 1990b. Deer mice as a case study in the operation of natural selection via differential reproductive success. Pages 129–148 in D.A. Dewsbury, ed., *Contemporary issues in comparative psychology.* Sinauer Associates, Sunderland, Massachusetts.

Dewsbury, D.A., and D.J. Baumgardner. 1981. Studies of sperm competition in two species of muroid rodents. *Behavioral Ecology and Sociobiology* 9:121–133.

Dewsbury, D.A., B. Ferguson, A.W. Hodges, and S.A. Taylor. 1986. Tests of preferences of deer mice (*Peromyscus maniculatus bairdi*) for individuals and their odors as a function of sex and estrous condition. *Journal of Comparative Psychology* 100:117–127.

Diakow, C. 1974. Male-female interactions and the organization of mammalian mating patterns. *Advances in the Study of Behavior* 5:227–268.

Dixon, K.R., and J.A. Chapman. 1980. Harmonic mean measure of animal activity areas. *Ecology* 61:1040–1044.

Doull, J. 1980. Factors influencing toxicology. Pages 70–83 in J. Doull, C.D. Klaassen, and M. O. Amdur, eds., *Casarett and Doull's toxicology.* Macmillan, New York.

Doyle, G.A., A. Pelletier, and T. Bekker. 1967. Courtship, mating and parturition in the lesser bushbaby (*Galago senegalensis moholi*) under semi-natural conditions. *Folia Primatologica* 7:169–197.

Draper, W.A., and I.S. Bernstein. 1963. Stereotyped behavior and cage size. *Perceptual and Motor Skills* 16:231–234.

Drickamer, L.C. 1974. A ten year summary of reproductive data for free-ranging *Macaca mulatta. Folia Primatologica* 21:60–80.

Driscoll, J.W, ed. 1989. *Animal care and use in behavioral research: Regulations, issues, and applications.* Animal Welfare Information Center, National Agricultural Library, U.S. Department of Agriculture, Beltsville, Maryland.

Drummond, B.A. 1994. Complex life histories and diverse reproductive physiologies of endangered invertebrates: implications for captive conservation. In E. Gibbons, B.S. Durrant and J. Demarest, eds.,

Conservation of endangered species in captivity: An interdisciplinary approach. State University of New York Press, Albany, New York.

Dryden, M.W. 1989. Biology of the cat flea *Ctenocephalides felis felis. Companion Animal Practice* 19:23–27.

Dubin, S., and S. Zietz. 1991. Sample size for animal health surveillance. *Lab Animal* 20:29–33.

Dubois, C., and M.K. Izard. 1990. Social and sexual behaviors in captive aye-ayes, *Daubentonia madagascariensis Journal of Psychological and Behavioral Science* 5:1–10.

Duffell, S.J., G.A.H. Wells, and C.E. Winkler. 1986. "Kangaroo gait" in ewes: a peripheral neuropathy. *Veterinary Record* 118:296–298.

Dufty, A.M., Jr., and J.C. Wingfield. 1986. The influence of social cues on the reproductive endocrinology of male brown-headed cowbirds: field and laboratory studies. *Hormones and Behavior* 20:222–234.

Duke-Elder, S., and P.A. MacFaul. 1972. Chorioretinal radiation burns. Pages 885–893 in S. Duke-Elder, ed., *System of opthalmology: Nonmechanical injuries.* C.V. Mosby, St. Louis.

Dunbar, R.I.M. 1976. Some aspects of research design and their implications in the observational study of behavior. *Behaviour* 58:78–98.

Duncan, I.J.H. 1981. Animal rights—animal welfare: a scientist's assessment. *Poultry Science* 60:489–499.

Durojaiye, O.A. 1984. Viral zoonoses in Nigeria: (2) non-rabies zoonoses. *International Journal of Zoonosis* 11:69–74.

Durrant, B.S., and K. Benirschke. 1981. Embryo transfer in exotic animals. *Theriogenology* 15:77–83.

Durrell, G. 1976. *The stationary ark.* Simon and Schuster, New York.

Duvall, S.W., I.S. Bernstein, and T.P. Gordon. 1976. Paternity and status in a rhesus monkey group. *Journal of Reproduction and Fertility* 47:25–31.

Eaton, R.L. 1974. *The cheetah: The biology, ecology and behavior of an endangered species.* Van Nostrand Reinhold, New York.

Ebeling, W. 1975. *Urban entomology.* University of California Press, Los Angeles.

Eibl-Eibesfeldt, I. 1961. The interactions of unlearned behaviour patterns and learning in mammals. Pages 53–73 in J.F. Delafresnaye, ed., *Brain Mechanisms and learning.* Thomas, Springfield.

————. 1970. *Ethology: The biology of behavior.* Holt, Rinehart and Winston, New York.

————. 1979. Human ethology: concepts and implications for the sciences of man. *The Behavioral and Brain Sciences* 2:1–57.

Eisenberg, J.F. 1963. The behavior of heteromyid rodents. *University of California Publications in Zoology* 69:1–114. University of California Press, Berkeley.

————. 1966. The social organizations of mammals. *Handbuch der Zoologie* Band 8, Lieferung 39, 10:1–92.

————. 1976. Communication mechanisms and social integration in the black spider monkey, *Ateles fusciceps robustus,* and related species. *Smithsonian Contributions to Zoology* 213:1–108.

Eisenberg, J.F., and D.G. Kleiman. 1977. The usefulness of behaviour studies in developing captive breeding programmes for mammals. *International Zoo Yearbook* 17:81–89.

Ellis, T.M. 1967. Environmental influences on drug responses in laboratory animals. Pages 569–588 in M.L. Conalty, ed., *Husbandry of laboratory animals.* Academic Press, London.

Elton, R.H., and B.V. Anderson. 1977. The social behavior of a group of baboons (*Papio anubis*) under artificial crowding. *Primates* 18:225–234.

Emery, D.E. 1986. Effects of endocrine state on sociosexual behavior of female rats tested in a complex environment. *Behavioral Neuroscience* 100:71–78.

Epple, G. 1978. Reproductive and social behavior of marmosets with special reference to captive breeding. Pages 50–62 in N. Gengozian and F. Deinhardt, eds., *Primates in medicine,* vol. 10, *Marmosets in experimental medicine.* S. Karger, Basel, Switzerland.

Eriksen, E. 1978. *Medikamentel immobilisering og indfangning af vilde kreaturer og hjortevildt.* Den hgl. Veterinaer-og Landbohojskoles, Copenhagen.

Erving III, G.W. 1985. Regulations and guidelines for animal care: problems and future concerns. Pages 281–296 in G.P. Moberg, ed., *Animal stress.* American Physiological Society, Bethesda, Maryland.

Erwin, J. 1979. Aggression in captive macaques: interaction of social and spatial factors. Pages 139–171 in J. Erwin, T. Maple, and G. Mitchell, eds., *Captivity and behavior: Primates in breeding colonies, laboratories, and zoos.* Van Nostrand Reinhold, New York.

Eterovic, V.A., and P.A. Ferchmin. 1985. Interaction with objects by group-living *Saimiri sciureus*. *Puerto Rico Health Sciences Journal* 4:121–125.

Evans, M., and W. Tempest. 1972. Some effects of infrasonic noise in transportation. *Journal of Sound and Vibration* 22:19–24.

Ewbank, R. 1985. Behavioral responses to stress in farm animals. Pages 71–79 in G.P. Moberg, ed., *Animal stress*. American Physiological Society, Bethesda, Maryland.

Fagen, R. 1981. *Animal play behavior*. Oxford Press, New York.

Falk, J. 1984. Excessive behavior and drug-taking: environmental generation and self-control. Pages 81–123 in P.K. Levison, ed., *Substance abuse, habitual behavior, and self-control*. Westview Press, Boulder, Colorado.

Faulkner, D.R. 1989. Design of a course to introduce research personnel to the care and use of laboratory animals. *Lab Animal* 18:21–25 .

Fisk, S.K., and W.H. Lewis. 1982. Animal facility design: designing with use & re-use in mind. *Lab Animal* 11: 38–39, 41.

Fleagle, J.G. 1979. Primate positional behavior and anatomy: naturalistic and experimental approaches. Pages 313–325 in M.E. Morbeck, H. Preuschoft, and N. Gomberg, eds., *Environment, behavior, and morphology: Dynamic interactions in primates*. Gustav Fischer, New York.

Fletcher, J.L. 1976. Influence of noise on animals. Pages 51–62 in T. McSheehy, ed., *Control of the animal house environment. Laboratory animal handbooks 7*. Laboratory Animals, London.

Flyger, V. F. 1960. Movements and home range of the gray squirrel, *Sciurus carolinensis*, in two Maryland woodlots. Ecology 41:363–369.

Flynn, R.J. 1959. Studies on the etiology of ringtail of rats. *Proceedings of the Animal Care Panel* 9 :155–160 .

———. 1968. A new cage cover as an aid to laboratory rodent disease control. *Proceedings of the Society for Experimental Biology and Medicine* 129:714–717.

Foerg, R. 1982. Reproductive behavior in *Varecia variegata*. *Folia Primatologica* 38:108–121.

Foerg, R., and R. Hoffman. 1982. Seasonal and daily activity changes in captive *Cheirogaleus medius*. *Folia Primatologica* 38:259–268 .

Folk, G. E., Jr. 1974. Biological rhythms. Pages 40–86 in G. E. Folk, Jr., *Textbook of environmental physiology*, 2nd. ed. Lea & Febiger, Philadelphia.

Folk, G.E., Jr., C.H. Dodge, and M.A. Folk. 1957. Resting body temperature of raccoons and domestic rabbits. *Proceedings of the Animal Care Panel* 7:253–258.

Follett, B.K., S. Ishii, and A. Chandola, eds. 1985. *The endocrine system and the environment*. Japan Scientific Societies Press, Tokyo, and Springer-Verlag, Berlin.

Forster, R.E. II, and T.B. Ferguson. 1952. Relationship between hypothalamic temperature and thermoregulatory effectors in unanesthetized cat. *American Journal of Physiology* 169:255–269.

Fox, J.G., B.J. Cohen, and F.M. Loew. 1984. *Laboratory animal medicine*. Academic Press, New York.

Fox, M.W., ed. 1968. *Abnormal behavior in animals*. W.B. Saunders, Philadelphia.

Fox, M.W. 1986. *Laboratory animal husbandry: Ethology, welfare, and experimental variables*. State University of New York Press, Albany, New York.

Fox, R. 1988. In the matter of "Baby M:" Report from the Gruter institute for law and behavioral research. *Politics and the Life Sciences* 7:77–85.

Fradrich, H. 1987. The husbandry of tropical and temperate cervids in the West Berlin zoo. Pages 422–428 in C.M. Wemmer, ed., *Biology and management of the Cervidae*. Smithsonian Institution Press, Washington, D.C.

Fragaszy, D.M., and W.A. Mason. 1978. Response to novelty in *Saimiri* and *Callicebus*: influence of social context. Primates 19:311–331.

Fragaszy, D. M., and E. Visalberghi. 1989. Social influences on the acquisition and use of tool-using behaviors in tufted capuchin monkeys (*Cebus apella*). *Journal of Comparative Psychology* 103:159–170.

Frank, W. 1984. Non-hemoparasitic Protozoonans. Pages 259–384 in G.L. Hoff, F.L. Frye and E.R. Jacobson, eds., *Diseases of amphibians and reptiles*. Plenum Press, New York.

Franks, P., and S. Lenington. 1986. Dominance and reproductive behavior of wild house mice in a seminatural environment correlated with T-locus genotype. *Behavioral Ecology and Sociobiology* 18:395–404.

Frederickson, W., and G. Sackett. 1984. Kin preferences in primates (*Macaca nemestrina*): relatedness or familiarity? *Journal of Comparative Psychology* 98:77–83 .

French, J.A., D.H. Abbott, G.A. Scheffler, J.A. Robinson, and R.W. Goy. 1983. Cyclic excretion of urinary oestrogens in female tamarins (*Saguinus oedipus*). *Journal of Reproduction and Fertility* 68:177–184.

French, J.A., D.H. Abbott, and C.T. Snowdon. 1984. The effect of social environment of estrogen excretion, scent marking, and sociosexual behavior in tamarins (*Saguinus oedipus*). *American Journal of Primatology* 6:155–167.

French, J.A., and B.J. Inglett. 1989. Female-female aggression and male indifference in responses to unfamiliar intruders in lion tamarins. *Animal Behaviour* 37:487–497.

French, J.A., and C.T. Snowdon. 1981. Sexual dimorphism in responses to unfamiliar intruders in the tamarin, *Saguinus oedipus*. *Animal Behaviour* 29:822–829 .

French, J.A., and J.A. Stribley. 1987. Synchronization of ovarian cycles within and between social groups in golden lion tamarins (*Leontopithecus rosalia*). *American Journal of Primatology* 12:469–478.

Friedmann, H. 1929. *The cowbirds: A study in the biology of social parasitism*. Charles C. Thomas, Springfield, Illinois.

Friedmann, H., L.F. Kiff, and S.I. Rothstein. 1977. A further contribution to knowledge of the host relations of the parasitic cowbirds. *Smithsonian Contributions to Zoology* 235:1–75.

Frings, H., and M. Frings. 1968. Practical uses. Pages 643–662 in T.A. Sebeok, ed. *Animal communication. Techniques of study and results of research*. Indiana University Press, Bloomington.

Fritschen, L.J., and L.W. Gay. 1979. *Environmental instrumentation*. Springer-Verlag, New York.

Fuller, J.L., and W.R. Thompson. 1960. *Behavior genetics*. John Wiley, New York.

Galbreath, G.J. 1982. Armadillo. Pages 71–79 in J.A. Chapman and G.A. Feldhamer, eds., *Wild mammals of North America: Biology, management, and economics*. Johns Hopkins University Press, Baltimore, Maryland.

Galef, B.G., Jr. 1982. Studies of social learning in Norway rats: a brief review. *Developmental Psychology* 15:279–295.

————. 1984. Reciprocal heuristics: a discussion of the relationship of learned behavior in laboratory and field. *Learning and Motivation* 15:479–493.

————. 1985. Direct and indirect behavioral pathways to the social transmission of food avoidance. *Annals of the New York Academy of Science* 443:203–215.

Galineo, S. 1964. Organ systems in adaptation: the temperature regulating system. Pages 259–282 in D.B. Dill, E.F. Adolph, and C.G. Wilber, eds., *Handbook of Physiology. Section 4: Adaptation to the environment.* American Physiological Society, Washington, D.C.

Galle, O.R., W.R. Gove, and J.M. McPherson. 1972. Population density and pathology: what are the relations for man? *Science* 176:23–30.

Gallistel, C.R. 1978. The use of animals in behavioral neurobiology: some ethical considerations. Paper presented at the 86th annual convention of the American Psychological Association, Toronto, Canada.

Ganzhorn, J.U. 1986. Feeding behavior of *Lemur catta* and *Lemur fulvus. International Journal of Primatology* 7:17–30.

Garber, P. 1984. Use of habitat and positional behavior in a Neotropical primate, *Saguinus oedipus.* Pages 112–133 in P. Rodman and J. Cant, eds., *Adaptations for Foraging in Nonhuman Primates.* Columbia University Press, New York.

Garcia, J. 1989. Food for Tolman: cognition and cathexis in concert. Pages 45–85 in T. Archer and L.-G. Nilsson, eds., *Aversion, avoidance, and anxiety: Perspectives on aversively motivated behavior.* Lawrence Erlbaum, Hillsdale, New Jersey.

Garretson, M.S. 1938. *The American bison.* New York Zoological Society, New York.

Garshelis, D.L., and M.R. Pelton. 1980. Activity of black bears in the Great Smoky Mountains National Park. *Journal of Mammalogy* 61:8–19.

Gautier, J.-P. 1988. Interspecific affinities among guenons as deduced from vocalizations. Pages 194–226 in A. Gautier-Hion, F. Bourliere, J.-P. Gautier and J. Kingdon, eds., *A primate radiation: Evolutionary biology of the African guenons.* Cambridge University Press, Cambridge, England.

Gautier, J.-P., and A. Gautier-Hion. 1982. Vocal communication within a group of monkeys: an analysis by biotelemetry. Pages 5–29 in C.T. Snowdon, C.H. Brown, and M.R. Petersen, eds., *Primate communication.* Cambridge University Press, Cambridge, England.

Gebo, D.L. 1987.   Locomotor diversity in prosimian primates. *American Journal of Primatology* 13:271–281.

Gengozian, N., J.S. Batson, and T.A. Smith. 1978.   Breeding of marmosets in a colony environment.   Pages 71–78 in N. Gengozian and F. Deinhardt, eds., *Primates in Medicine*, vol. 10: *Marmosets in Experimental Medicine.* S. Karger, Basel, Switzerland.

Gettinger, R.D. 1975.   Metabolism and thermoregulation of a fossorial rodent, the Northern pocket gopher (*Thomomys talpoides*).   *Physiological Zoology* 48:311–322.

Getz, L.L. 1978.   Speculation on social structure and population cycles of microtine rodents.   *The Biologist* 60:134–147.

Getz, L.L., and J.E. Hofmann. 1986.   Social organization in free-living prairie voles, *Microtus ochrogaster.   Behavioral Ecology and Sociobiology* 18:275–282.

Gibbons, E.F., Jr. 1994. Conservation of endangered primates in captivity. In E. Gibbons, B.S. Durrant and J. Demarest, eds., *Conservation of endangered and threatened species in captivity: An interdisciplinary approach.*   State University of New York Press, Albany, New York.

Gibbons, E.F., Jr., and B.S. Durrant. 1987.   Behavior and development in offspring from interspecies embryo transfer: theoretical issues. *Applied Animal Behaviour Science* 18:105–118.

Gibbons, E.F., Jr., B.S. Durrant and J. Demarest, eds. 1994. *Conservation of endangered species in captivity: An interdisciplinary approach.* State University of New York Press, Albany, New York.

Gibbons, E.F., Jr., F. Koontz, R. Cook, and M.K. Stoskopf. 1992. Behavioral medicine in zoological parks and aquariums.   Paper presented at the Thirteenth Annual Scientific Sessions of the Society of Behavioral Medicine, New York.

Gibbons, E.F., Jr., and M.K. Stoskopf. 1989.   An interdisciplinary approach to animal medical problems.   Pages 60–68 in J. Driscoll, ed., *Animal care and use in behavioral research: Regulations, issues and applications.*   Animal Welfare Information Center, U.S. Department of Agriculture, National Agricultural Library, Beltsville, Maryland.

Gibson, E.J. 1969.   *Principles of perceptual learning and development.* Appleton-Century-Crofts, New York.

Gibson, E.J., and R.D. Walk. 1956. The effect of prolonged exposure to visually presented patterns on learning to discriminate them. *Journal of Comparative and Physiological Psychology* 49:239–242.

Gibson, J.J. 1979. *The ecological approach to visual perception.* Houghton Mifflin, Boston.

Ginsburg, B.E. 1979. The violent brain: is it everyone's brain? Pages 47–64 in C.R. Jeffery, ed., Biology and Crime. *Sage Research Progress Series in Criminology, Volume 10.* Sage Publications (in cooperation with the American Society of Criminology), Beverly Hills, California.

Ginsburg, B.E. 1987. The wolf pack as a socio-genetic unit. Pages 401–413 in H. Frank, ed., *Man and wolf: Advances, issues and problems in captive wolf research.* Dr. W. Junk, Dordrecht, The Netherlands.

Ginsburg, B.E., and W.C. Allee. 1942. Some effects of conditioning on social dominance and subordination in inbred strains of mice. *Physiological Zoology* 15:485–506.

Ginsburg, B.E., and C.S. Schotte. 1979. The effects of visual isolation on a captive wolf pack. *Animal Behavior Society Abstracts*, no. 68.

Gittleman, J.L., and G.K. Conover. 1994. Mammalian behavior: lessons from captive studies. In E. Gibbons, B.S. Durrant and J. Demarest, eds., *Conservation of endangered species in captivity: An interdisciplinary approach.* State University of New York Press, Albany, New York.

Glander, K.E. 1978. Howling monkey feeding behavior and plant secondary compounds: a study of strategies. Pages 561–573 in G.G. Montgomery, ed., *The ecology of arboreal folivores.* Smithsonian Institution Press, Washington, D.C.

Glander, K.E., and D.P. Rabin. 1983. Food choice from endemic North Carolina tree species by captive prosimians (*Lemur fulvus*). *American Journal of Primatology* 5:221–529.

Glickman, S.E. 1973. Responses and reinforcement. Pages 207–241 in R.A. Hinde and J.S. Hinde, eds., *Constraints on learning: Limitation and predispositions.* Academic Press, London.

Glickman, S.E., L. Fried, and B. Morrison. 1967. Shredding of nesting material in the Mongolian gerbil. *Perceptual and Motor Skills* 24:473–474.

Glickman, S.E., K. Holekamp, and B. Ivins. 1982. Seasonality, reproduction and "house" occupancy in a population of woodrats. Paper presented at the annual meeting of the Animal Behavior Society, Duluth, Minnesota.

Glickman, S.E., and R.W. Sroges. 1966. Curiosity in zoo animals. *Behaviour* 26:151–188.

Goerke, B., L. Fleming, and M. Creel. 1987. Behavioral changes of a juvenile gorilla after a transfer to a more naturalistic environment. *Zoo Biology* 6:283–295.

Goethe, F. 1937. Beobachtungen und untersuchungen zur biologie der silbermowe (*Larus a. argentatus Pontopp.*) auf der vogelinsel Memmertsand. *Journal für Ornithologie* 85:1–119.

Gold, R.E. 1985. Cockroach resistance to insecticides as it relates to the current state-of-the-art in population management. *Proceedings of the 1st Insect Growth Regulator Symposium* 1:40–47 .

Goldstein, I. S. 1988. Measurement and characteristics of light. Pages 31–54 in D. C. Holley, C.M. Winget and H.A. Leon, eds., *Lighting requirements in microgravity-rodents and nonhuman primates.* National Aeronautics and Space Administration, Ames Rescarch Center, Moffett Field, California.

Goldstein, J. 1978. Fundamental concepts in sound measurement. Pages 3–58 in D.M. Lipscomb, ed., *Noise and audiology.* University Park Press, Baltimore.

Goldstein, S.J. 1981. Disturbed-site feeding by rhesus monkeys in Pakistan. *American Journal of Physical Anthropology* 54:225–226 (abstract).

Goldstein, S.R. 1981. An operant arena for rats. *Behavior Research Methods & Instrumentation* 13: 37–39 .

Gonyou, H.W. 1986. Assessment of comfort and well-being in farm animals. *Journal of Animal Science* 62:1769–1775.

Goodall, J. 1986. *The chimpanzees of Gombe: Patterns of behavior.* Belknap Press, Cambridge, Massachusetts.

Goosen, C. 1981. Abnormal behavior patterns in rhesus monkeys: symptoms of mental disease? *Biological Psychiatry* 16: 697–716 .

————. 1988. Developing housing facilities for rhesus monkeys: prevention of abnormal behavior. Pages 59–70 in A. C. Beynen and H.A. Solleveld, eds., *New developments in biosciences: Their implications for laboratory animal science.* Martinus Nijhoff, Dordrecht, The Netherlands.

Gould, S.J., and R.C. Lewontin. 1978. The spandrels of San Marco and the Panglossian paradigm: a critique of the adaptationist program. *Proceedings of the Royal Society of London,* Serial B, 205:591–598.

Graves, R.G. 1990. Animal facilities: planning for flexibility. *Lab Animal* 19:29–50.

Greenough, W.T. 1975. Experiential modification of the developing brain. *American Scientist* 63:37–46.

Grothaus, R.J., D.E. Weidhaas, D.G. Haile, and G.S. Burden. 1981. Super-roach challenged by microcomputer. *Pest Control* 49:16–18.

Gut, I., M. Cikrt, and G.L. Plaa, eds. 1981. *Industrial and environmental xenobiotics.* Springer-Verlag, Berlin.

Gwosdow, A.R., and E.L. Besch. 1985. Effect of thermal history on the rat's response to varying environmental temperature. *Journal of Applied Physiology* 59:413–419.

Hafez, E.S.E. 1968. Principles of animal adaptation. Pages 3–17 in E.S.E. Hafez, ed., *Adaptation of domestic animals.* Lea & Febiger, Philadelphia.

Hale, E.B. 1969. Domestication and the evolution of behavior. Pages 22–44 in E.S.E. Hafez, ed., *The behaviour of domestic animals*, 2nd. ed. Bailliere, Tindall, and Cassell, London.

Hamilton, W.J., III, and P.C. Arrowood. 1978. Copulatory vocalizations of chacma baboons (*Papio ursinus*), gibbons (*Hylobates hoolock*), and humans. *Science* 200:1405–1409.

Hammel, H.T, C.H. Wyndham, and J.D. Hardy. 1958. Heat production and heat loss in the dog at 8–36°C environmental temperature. *American Journal of Physiology* 194:99–108.

Hancocks. D. 1971. *Animals and architecture.* Praeger, New York.

———. 1980. Bringing nature into the zoo: inexpensive solutions for zoo environments. *International Journal for the Study of Animal Problems* 1:170–177.

Hansen, T.E., and D.W. Mangold. 1960. Functional and basic requirements of swine housing. *Agricultural Engineering* 41:585–590.

Hanson, J.D., M.E. Larson, and C.T. Snowdon. 1976. The effects of control over high intensity noise on plasma cortisol levels in rhesus monkeys. *Behavioral Biology* 16:333–340.

Hanson, L.E. 1982. Leptospirosis in domestic animals: the public health perspective. *Journal of the American Veterinary Medical Association* 181:1505–1509.

Harcourt, C.S., and L.T. Nash. 1986. Social organization of galagos in Kenyan coastal forests: I. *Galago zanzibaricus. American Journal of Primatology* 10:339–355.

Harlow, H.F. 1949. The formation of learning sets. *Psychological Review* 56:51–56.

Harlow, H.F., and M.K. Harlow. 1962. Social deprivation in monkeys. *Scientific American* 207(5):136–146.

Harlow, H.F., and M.K. Harlow. 1969. Effects of various mother-infant relationships on rhesus monkey behaviors. Pages 15–36 in B.M. Foss, ed., *Determinants of infant behavior*, volume 4. Methuen, London.

Harlow, H.F., and S.J. Soumi. 1971. Social recovery by isolation reared monkeys. *Proceedings of the National Academy of Sciences, U.S.A.* 68 :1534–1538.

Harrington, J. E. 1976 . Discrimination between individuals by scent in *Lemur fulvus. Animal Behaviour* 24:207–212.

———. 1977. Discrimination between males and females by scent in *Lemur fulvus. Animal Behaviour* 25:147–151.

Harris, V.T. 1952. An experimental study of habitat selection by prairie and forest races of the deermouse, *Peromyscus maniculatus. Contributions from the Laboratory of Vertebrate Biology, University of Michigan, Ann Arbor*, no. 56.

Harrison, L.P. 1965. Fundamental concepts and definitions relating to humidity. Pages 3–69 in A. Wexler and W.A. Wildhuck, eds., *Humidity and moisture: Measurement and control in science and industry*, volume 3. *Fundamentals and standards.* Reinhold, New York.

Harthoorn, A.M. 1976. *The chemical capture of animals.* Bailliere, Tindall, London.

Hartung, T.G., and D.A. Dewsbury. 1979. Paternal behavior in six species of muroid rodents. *Behavioral and Neural Biology* 26:466–478.

Hastings, J. S . 1967. Long term use of vermiculite. *Journal of the Institute of Animal Technicans* 18:184–190.

Hastings, J.W., and M. Menaker, Chairmen. 1976. Physiological and biochemical aspects of circadian rhythms. *Federation Proceedings* 35:2325–2357.

Hausfater, G. 1975. Dominance and reproduction in baboons (*Papio cynocephalus*). *Contributions to Primatology*, volume 7. S. Karger, Basel, Switzerland.

Hayden, C.C. 1987. How to increase employee participation in the lab animal environment. *Lab Animal* 16:47–49.

Hayes, S.L., and C.T. Snowdon. 1990. Predator recognition in cotton-top tamarins (*Saguinus oedipus*). *American Journal of Primatology* 20:283–291.

Hays, W.L. 1973. *Statistics for the social sciences.* Holt, Rinehart and Winston, New York.

Hazen, T.E., and D.W. Mangold. 1960. Functional and basic requirements of swine housing. *Agricultural Engineering* 41:585–590.

Hazlett, B.A. ed. 1977. Quantitative methods in the study of animal behaviour. *Academic Press,* New York.

Hediger, H. 1955. *The psychology and behavior of captive animals in zoos and circuses.* Dover, New York.

———. 1964. *Wild animals in captivity.* Dover Publications, New York.

———. 1965. Environmental factors affecting reproduction of zoo animals. Pages 319–354 in F.A. Beach, ed., *Sex and behavior.* John Wiley, New York.

———. 1969. *Man and animal in the zoo.* Delacorte Press, New York.

Heiser, C.B., Jr. 1990. *Seed to civilization.* Harvard University Press, Cambridge.

Heller, E. 1930. The american prong-horned antelope (*Antilocapra americana*) with special reference to the history of a pair and their progeny at the Milwaukee Zoological Garden. *Bulletin of the Washington Park Zoological Society* 1:1–8.

Herrington, L.P. 1940. The heat regulation of small laboratory animals at various environmental temperatures. *American Journal of Physiology* 129:123–139.

Heske, E.J., and R.J. Nelson. 1984. Pregnancy interruption in *Microtus ochrogaster*: laboratory artifact or field phenomenon? *Biology of Reproduction* 31:97–103.

Hess, E.H., S.B. Petrovich, and E.B. Goodwin. 1976. Induction of parental behavior in Japanese quail (*Coturnix coturnix japonica*). *Journal of Comparative and Physiological Psychology* 90:244–251.

Hill, J.L. 1987. The rat populations of NIMH: density, reproduction, and the neocortex. *American Zoologist* 27:839–851.

Hill, W.F. 1978. Effects of mere exposure on preferences in nonhuman mammals. *Psychological Bulletin* 85:1177–1198.

Hinde, R.A 1973.   On the design of check-sheets. *Primates* 14:393–406.

Hinde, R.H. 1974.   *Biological bases of human social behavior.* McGraw-Hill, New York.

Hoage, R.J. 1977.   *Parental care in Leontopithecus rosalia rosalia:* sex and age differences in carrying behavior and the role of prior experience. Pages 293–305 in D.G. Kleiman, ed., *The biology and conservation of the Callitrichidae.* Smithsonian Institution Press, Washington, D.C.

Hodgson, E. 1980.   Chemical and environmental factors affecting metabolism of xenobiotics. Pages 143–161 in E. Hodgson and F. E. Guthrie, eds., *Introduction to Biochemical Toxicology.* Elsevier, New York.

Hoffman, R., and R. Foerg. 1983. Seasonal and daily activity changes in captive *Cheiroaleus medius. Folia Primatologica* 38:259–268.

Hogan, J., and T.J. Roper. 1978. A comparison of the properties of different reinforcers. Pages 156–255 in J.S. Rosenblatt, R.A. Hinde, C. Beer and M.C. Busnel eds., *Advances in the study of behavior.* Academic Press, New York.

Holmes, W., and P. Sherman. 1982.   The ontogeny of kin recognition in two species of ground squirrels. *American Zoologist* 22:491–517.

Hooppes, P.J., and R.J. Montali. 1980.   Tail lesions in captive elephant shrews.   Pages 425–430 in R.J. Montali and G. Migaki, eds., *The comparative pathology of zoo animals.* Smithsonian Institution Press, Washington, D.C.

Horsfall, W.R. 1985. Mosquito control in a changing world. *Journal of the American Mosquito Control Association* 1:135–138.

Howard, W.E., and H.E. Childs, Jr. 1959. Ecology of pocket gophers with emphasis on *Thomomys bottae mewa. Hilgardia* 29:277–358.

Huff, D. 1954. *How to lie with statistics.* Norton, New York.

Hughes, L., M.K. Izard, M.A. Morris, and D.L. Hatchell. 1990. Diabetes mellitus not associated with late pregnancy wastage in prosimian primates. *Advances in Contraceptive and Delivery Systems* 5:245–249.

Hurst, J.L. 1986. Mating in free-living wild house mice (*Mus domesticus*). *Journal of Zoology* 210:623–628.

Hurst, J.L. 1987. Behavioural variation in wild house mice *Mus domesticus* Rutty: a quantitative assessment of female social organization. *Animal Behaviour* 35:1846–1857.

Hutchins, M. 1988. On the design of zoo research programmes. *International Zoo Yearbook* 27:9–19.

Hutchins, M., D. Hancocks, and C. Crockett. 1978. Naturalistic solutions to the behavioral problems of captive animals. *American Association of Zoological Parks and Aquariums Conference Proceedings*, 108–113.

Hutchins, M., C. Sheppard, A.M. Lyles, and G. Casadei. 1994. Behavioral considerations in the captive management, propagation and reintroduction of endangered birds. In E. Gibbons, B.S. Durrant and J. Demarest, eds., *Conservation of endangered species in captivity: An interdisciplinary approach.* State University of New York Press, Albany, New York.

Hutt, S.J., and C. Hutt. 1970. *Direct observation and measurement of behavior.* Charles C. Thomas, Springfield, Illinois.

Ingles, L.G. 1965. *Mammals of the pacific states: California, Oregon, and Washington.* Stanford University Press, Stanford, California.

Inhelder, E. 1962. Skizzen zu einer Verhältenspathologie reactiver Störungen bei Tieren. *Schweizer Archiv für Neurologie, Neurochirurgie und Psychiatrie* 89:276–326.

Institute of Laboratory Animal Resources Committee on Cats. 1978. Laboratory animal management—cats. Institute of *Laboratory Animal Resources News* 21(3):C1-C20.

Institute of Laboratory Animal Resources Committee on Rodents. 1977. Laboratory animal management—rodents. *Institute of Laboratory Animal Resources News* 20:Ll–L15.

Institute of Laboratory Animal Resources Committee on Standards. 1965. *Standards for the breeding, care, and management of laboratory animals: Rabbits.* National Academy of Sciences, Washington, D.C.

Institute of Laboratory Animal Resources Subcommittee on Avian Standards, Committee on Standards. 1966. *Standards and guidelines for the breeding, care and management of laboratory animals: Chickens.* National Academy of Sciences, Washington, D.C.

Institute of Laboratory Animal Resources Subcommittee on Care and Use, Committee on Nonhuman Primates. 1980. Laboratory animal management: Nonhuman primate. *Institute of Laboratory Resources News* 23:P1-P44.

Institute of Laboratory Animal Resources Committee on Care and Use of Laboratory Animals. 1985. *Guide for the care and use of laboratory animals*, NIH publication 86–23. U.S. Department of Health and Human Services, Washington, D.C.

Institute of Laboratory Animal Resources Subcommittee on Dog and Cat Standards, Committee on Standards. 1973a. *Dogs: Standards and guidelines for the breeding, care, and management of laboratory animals*. National Academy of Sciences, Washington, D.C.

Institute of Laboratory Animal Resources Subcommittee on Revision of Non-human Primate Standards, Committee on Standards. 1973b. *Nonhuman Primates: Standards and guidelines for the breeding, care, and management of laboratory animals*. National Academy of Sciences, Washington, D.C.

International Union for Conservation of Nature and Natural Resources. 1988. *IUCN Red List of Threatened Animals*. International Union for Conservation of Nature and Natural Resources, Gland, Switzerland.

Ishii, S., and Y. Kuwahara. 1968. Aggregation of German cockroach (*Blattella germanica*) nymphs. *Experientia* 24:88–89.

Ising, H. 1981. Interaction of noise-induced stress and Mg decrease. *Artery* 9:205–211.

Izard, M.K. 1989. Duke University primate center. *Journal of Medical Primatology* 18:85–98.

Izard, M.K., and P.A. Fail. 1988. Progesterone levels during pregnancy in the greater thick-tailed galago (*Galago crassicaudatus*). *Journal of Medical Primatology* 17:125–133.

Izard, M.K., S.J. Heath, Y. Hayes, and E.L. Simons. 1991. Hematology, serum chemistry values, and rectal temperatures of adult greater galagos (*Galago garnettii* and *G. crassicaudatus*). *Journal of Medical Primatology* 20:117–121.

Izard, M.K., and D.T. Rasmussen. 1985. Reproduction in the slender loris (*Loris tardigradus malabaricus*). *American Journal of Primatology* 8:153–165.

Izard, M.K., and E.L. Simons. 1986. Management of reproduction in a breeding colony of bushbabies. Pages 315–323 in J.G. Else and P.C. Lee, eds., *Primate ecology and conservation*. Cambridge University Press, Cambridge, England.

Izawa, K. 1979. Foods and feeding behavior of wild black-capped capuchins (*Cebus apella*). Primates 20:57–76.

Izawa, K., and A. Mizuno. 1977. Palm-fruit cracking behavior of wild black-capped capuchins (*Cebus apella*). Primates 18:773–792.

Janson, C.H. 1984. Female choice and mating system of the brown capuchin monkey *Cebus apella* (Primates: Cebidae). *Zeitschrift für Tierpsychologie* 65:177–200.

———. 1985. Aggressive competition and individual food consumption in wild brown capuchin monkeys (*Cebus apella*). *Behavioral Ecology and Sociobiology* 18:125–138.

———. 1988a. Intra-specific food competition and primate social structure: a synthesis. *Behaviour* 105:1–17.

———. 1988b. Food competition in brown capuchin monkeys (*Cebus apella*): quantitative effects of group size and tree productivity. *Behaviour* 105:53–76.

———. 1990. Ecological consequences of individual spatial choice in foraging groups of brown capuchin monkeys, *Cebus apella*. *Animal Behaviour* 40:922–934.

Janson, C.H., E.W. Stiles, and D.W. White. 1986. Selection on plant fruiting traits by brown capuchin monkeys: a multivariate approach. Pages 83–92 in A. Estrada and T.H. Fleming, eds., *Frugivores and seed dispersal*. D.W. Junk, Dordrecht, The Netherlands.

Johnson, E., M. Miller, and R.H. Smith. 1980. Radio telemetry of heart rate and temperature in fallow deer. Pages 563–567 in C.J. Amlaner and D.W. MacDonald, eds., *A handbook on biotelemetry and radio tracking*. Pergamon Press, Oxford.

Johnson, G.S., and R.S. Elizondo. 1979. Thermoregulation in *Macaca mulatta*: a thermal balance study. *Journal of Applied Physiology: Respiration, Environment, Exercise Physiology* 46:268–277.

Johnson, H.D., and W.J. Vanjonack. 1976. Effects of environmental and other stressors in blood hormone patterns in lactating animals. *Journal of Dairy Science* 59:1603–1617.

Johnsson, T., J.F. Lavender, and J.T. Marsh. 1959. The influence of avoidance-learning stress on resistance to Coxsackie virus in mice. *Federation Proceedings* 18:575 (abstract).

Johnston, T.J. 1988. Developmental explanation and the ontogeny of birdsong: nature-nurture-redux. *Behavioral and Brain Sciences* 11:617–663.

Jolly, A. 1986. Lemur survival. Pages 71–98 in K. Benirschke, ed., *Primates: The road to self-sustaining populations.* Springer-Verlag, New York.

Jori, A., A. Bianchetti, and P.E. Prestini. 1969. Effect of essential oils on drug metabolism. *Biochemical Pharmarcology* 18:2081–2085.

Jouffroy, F.K., and J. Lessertisseur. 1979. Some comments on the methodological approach to the interface—morphology, behavior, environment. Pages 23–31 in M.E. Morbeck, H. Preuschoft, and N. Gomberg, eds., *Environment, behavior, and morphology: Dynamic interactions in primates.* Gustav Fischer, New York.

Jurtshak, P., A.S. Weltman, and A.M. Sackler. 1959. Biochemical responses of rats to auditory stress. *Science* 129:1424–1425.

Kamil, A.C., S.I. Yoerg, and K.C. Clements. 1988. Rules to leave by: patch departure in foraging blue jays. *Animal Behaviour* 36:843–853.

Kang, B., and C. Morgan. 1980. Incidence of allergic skin reactivities of asthmatics to inhalant allergens. *Clinical Research* 28:426A (abstract).

Kaplan, J.R. 1986. Psychological stress and behavior in nonhuman primates. Pages 455–92 in G. Mitchell and J. Erwin, eds., *Comparative primate biology*, volume 2A: *Behavior, conservation, and ecology.* Alan R. Liss, New York.

Kappeler, P.M. 1987a. The acquisition process of a novel behavior pattern in a group of ring-tailed lemurs (*Lemur catta*). *Primates* 28:225–228.

———. 1987b. Reproduction in the crowned lemur (*Lemur coronatus*) in captivity. *American Journal of Primatology* 12:497–503.

———. 1990. Female dominance in *Lemur catta*: more than just feeding priority? *Folia Primatologica* 55:92–95.

Katz, A.S. 1980. Management techniques to reduce perinatal loss in a lemur colony. *American Association of Zoological Parks and aquariums Regional Proceedings* 137–140.

Kavanau, J.L. 1963. Compulsory regime and control of environment in animal behavior. I. Wheel-running. *Behaviour* 20:251–281.

———. 1964. Behavior: confinement, adaptation, and compulsory regimes in laboratory studies. *Science* 143:490.

———. 1967a. Wheel-running preferences of mice. *Zietschrift für Tierpsychologie* 24:858–866.

————. 1967b. Behavior of captive white-footed mice. *Science* 155:1623–1639.

Kavanau, J.L., and D.H. Brant. 1965. Wheel-running preferences of *Peromyscus*. *Nature* 208:597–598.

Kearton, C. 1946. *In the land of the lion*. Arrowsmith, London.

Keast, D., and M.F. Coales. 1967. Lymphocytopenia induced in a strain of laboratory mice by agents commonly used in treatment of ectoparasites. *Australian Journal of Experimental Biology and Medical Science* 45:645–650.

Keeling, M.E., P.L. Alford, and M.A. Bloomstrand. 1988. Psychological well-being of semi-free ranging chimpanzees. *American Journal of Primatology* 14:428 (abstract).

Keitz, H.A.E. 1971. *Light calculations and measurements: An introduction to the system of quantities and units in light-technology and to photometry*. St. Martin's Press, New York.

Kelley, K.W. 1985. Immunological consequences of changing environmental stimuli. Pages 193–223 in G.P. Moberg, ed., *Animal stress*. American Physiological Society, Bethesda, Maryland.

Kenshalo, R. 1990. Correlations of temperature sensation and neural activity: a second approximation. Pages 67–88 in J. Bligh and K. Voigt, eds., *Thermoreception and temperature regulation*. Springer-Verlag, Berlin.

Keplinger, M.L., G.E. Lanier, and W.8. Deichmann. 1959. Effects of environmental temperature on the acute toxicity of a number of compounds in rats. *Toxicology and Applied Pharmacology:* 1:156–161.

Kinder, E.F. 1927. A study of the nest-building activity of the albino rat. *Journal of Experimental Zoology* 47:117–161.

King, A.P., and M.J. West. 1977. Species identification in the North American cowbird: appropriate responses to abnormal song. *Science* 195:1002–1004.

King, A.P., M.J. West, and D.H. Eastzer. 1980. Song structure and song development as potential contributors to reproductive isolation in cowbirds (*Molothrus ater*). *Journal of Comparative and Physiological Psychology* 94:1028–1039.

King, A.P., and M.J. West. 1983a. Dissecting cowbird song potency: assessing a song's geographic identity and relative appeal. *Zeitschrift für Tierpsychologie* 63:37–50.

King, A.P., and M.J. West. 1983b. Female perception of cowbird song: a closed developmental program. *Developmental Psychobiology* 16:335–342.

King, A.P., and M.J. West. 1983c. Epigenesis of cowbird song—a joint endeavor of males and females. *Nature* 305:704–706.

King, A.P., and M.J. West. 1988. Searching for the functional origins of song in eastern brown-headed cowbirds, *Molothrus ater ater*. *Animal Behaviour* 36:1575–1588.

King, A.P., and M.J. West. 1989. Presence of female cowbirds (*Molothrus ater ater*) affects vocal imitation and improvisation in males. *Journal of Comparative Psychology* 103:39–44.

King, M.-C., and A.C. Wilson. 1975. Evolution at two levels in humans and chimpanzees. *Science* 188:107–116.

Kleiman, D.G. 1981. *Leontopithecus rosalia*. Mammalian Species, Provo, Utah. *The American Society of Mammalogists* (Brigham Young University), no. 148:1–7.

Kleiman, D. 1989. Reintroduction of captive mammals for conservation: guidelines for reintroducing endangered species into the wild. *BioScience* 39:152–161.

Kleiman, D., B. Beck, J. Dietz, L. Dietz, J. Ballou, and A. Coimbra-Filho. 1986. Conservation program for the golden lion tamarin: captive research and management, ecological studies, educational strategies, and reintroduction. Pages 959–979 in K. Benirschke, ed., *Primates: The road to self-sustaining populations*. Springer-Verlag, New York.

Kleiman, D., B. Beck, J. Dietz, and L. Dietz. 1991. Costs of a reintroduction and criteria for success: accounting and accountability in the Golden Lion Tamarin Conservation Program. In J. Gipps, ed., *Beyond captive breeding: Reintroducing endangered species to the wild. Symposia of the Zoological Society of London* 62:125–142.

Kline, L.W. 1898. Methods in animal psychology. *American Journal of Psychology* 10:256–279.

Kling, H.F., and C.L. Quarles. 1974. Effect of atmospheric ammonia and the stress of infectious bronchitis vaccination on leghorn males. *Poultry Science* 53:1161–1167.

Koluchova, J. 1972. Severe deprivation in twins: a case study. *Journal of Child Psychology and Psychiatry* 13:107–114.

————. 1976. A report on further development of twins after severe and prolonged deprivation. Pages 56–66 in A.M. Clarke and A.D.B. Clarke, eds., *Early experience: Myth and evidence.* Free Press, New York.

Konstant, W.R., and R.A. Mittermeier. 1982. Introduction, reintroduction and translocation of Neotropical primates: past experiences and future possibilities. *International Zoo Yearbook* 22:69–77.

Koontz, F.W., and P.R. Thomas. 1989. Applied ethology as a tool for improving animal care in zoos. Pages 69–80 in J.W. Driscoll, ed., *Animal care and use in behavioral research: Regulations, issues, and applications.* Animal Welfare Information Center, National Agricultural Library, Beltsville, Maryland.

Krebs, J.R., and R.H. McCleery. 1984. Optimization in behavioral ecology. Pages 91–121 in J.R. Krebs and N.B. Davies, eds., *Behavioral ecology: An evolutionary approach*, 2nd ed. Sinauer Associates, Sunderland, Massachusetts.

Krechevsky, I. 1938. An experimental investigation of the principle of proximity in the visual perception of the rat. *Journal of Experimental Psychology* 22:497–523.

Krog, H., and M. Monson. 1954. Notes on the metabolism of a mountain goat. *American Journal of Physiology* 178:515–516.

Kroll, J.L., and P.L. Johnson. 1980. A computer based file management system for telemetry studies on free ranging animals. Pages 425–430 in C.J. Amlaner and D.W. MacDonald, eds., *A handbook on biotelemetry and radio tracking.* Pergamon Press, Oxford.

Kroodsma, D.E., and J.R. Baylis. 1982. Appendix: a world survey of evidence for vocal learning in birds. Pages 311–337 in D.E. Kroodsma and E.H. Miller, eds., *Acoustic communication in birds*, volume 2. *Song learning and its consequences.* Academic Press, New York.

Krusko, N., M. Weldele, and S.E. Glickman. 1988. Meeting ceremonies in spotted hyenas. Poster presented at the annual meeting of the Animal Behavior Society, Missoula, Montana.

Kruuk, H. 1972. *The spotted hyena: A study of predation and social behavior.* University of Chicago Press, Chicago.

Kummer, H. 1982. Social knowledge in free-ranging primates. Pages 113–130 in D. Griffin, ed., *Animal mind—human mind.* Springer-Verlag, Berlin.

Kuo, Z.Y. 1969. *The dynamics of behavior development: An epigenetic view.* Random House, New York.

Lane-Petter, W. 1963. The physical environment of rats and mice. Pages 1–20 in W. Lane-Petter, ed., *Animals for research: Principles of breeding and management.* Academic Press, London.

Lang, C.M., and E.S. Vesell. 1976. Environmental and genetic factors affecting laboratory animals: impact on biomedical research. *Federation Proceedings* 35:1123–1124.

Larter, R.J., and P.R. Chadwick. 1983. Use of a general model to examine control procedures for a cockroach population. *Researches on Population Ecology* 25:238–248.

Lasley, B.L. 1985. Methods for evaluating reproductive function in exotic species. *Advances in Veterinary Science and Comparative Medicine* 30:209–228.

Lee, R.C. 1939. Basal metabolism of the adult rabbit and prerequisites for its measurement. *Journal of Nutrition* 18:473–488.

Lee, R.C., N.F. Colovos, and E.G. Ritzman. 1941. Skin temperatures of the pig, goat, and sheep under winter onditions. *Journal of Nutrition* 21:321–326.

Lefcourt, H.M. 1973. The function of the illusions of control and freedom. *American Psychologist* 28:417–425.

Lehner, P.N. 1979. *Handbook of ethological methods.* Garland STPM Press, New York.

Lehrman, D.S. 1953. A critique of Konrad Lorenz's theory of instinctive behavior. *Quarterly Review of Biology* 28:337–363.

Leopold, A. 1986. *Game management.* University of Wisconsin Press, Madison.

Levine, S. 1983. Coping: an overview. Pages 15–26 in H. Ursin and R. Murison, eds., *Biological and psychological basis of psychosomatic disease.* Pergamon Press, New York.

———. 1985. A definition of stress? Pages 51–69 in G. P. Moberg, ed., *Animal stress.* American Physiological Society, Bethesda, Maryland.

Lickliter, R., and G. Gottlieb. 1985. Social interaction with siblings is necessary for visual imprinting of species-specific maternal preferences in ducklings (*Anas platyrhynchos*). *Journal of Comparative Psychology* 99:371–379.

Lindburg, D.G. 1971. The rhesus monkey in North India: an ecological and behavioral study. Pages 1–106 in L.A. Rosenblum, ed., *Primate behavior: Developments in field and laboratory research,* volume 2. Academic Press, New York.

———. 1988. Improving the feeding of captive felines through application of field data. *Zoo Biology* 7:211–218.

Lindsey, J.R., M.W. Conner, and H.J. Baker. 1978. Physical, chemical, and microbiol factors affecting biologic response. Pages 31–43 in E. Besch, Chairman, *Laboratory animal housing.* National Academy of Sciences, Institute of Laboratory Animal Resources, Washington, D.C.

Line, S.W. 1987. Environmental enrichment for laboratory primates. *Journal of the American Veterinary Medical Association* 190:854–859.

Line, S.W., K.N. Morgan, and H. Markowitz. 1991. Simple toys do not alter the behavior of aged rhesus monkeys. *Zoo Biology* 10:473–484.

Linc, S.W., K.N. Morgan, H. Markowitz, and S. Strong. 1990. Increased cage size does not alter heart rate or behavior in female rhesus monkeys. *American Journal of Primatology* 20:107–114.

Linsdaile, J.M., and L.P. Tevis, Jr. 1951. *The dusky-footed wood rat.* University of California Press, Berkeley.

Lisk, R.D., U.W. Huck, A.C. Gore, and M.X. Armstrong. 1989. Mate choice, mate guarding and other mating tactics in golden hamsters maintained under seminatural conditions. *Behaviour* 109:58–75.

Lofgreen, P.E., Jr. 1987. Computers in the animal facility. *Lab Animal* 16:59–63.

Lord, R.D., Jr. 1961. A population study of the gray fox. *American Midland Naturalist* 66:87–109.

Lore, R. and K. Flannelly. 1977. *Rat societies. Scientific American* 236:106–116.

Lore, R., M. Nikoletseas, and L. Takahashi. 1984. Colony aggression in laboratory rats: a review and some recommendations. *Aggressive Behavior* 10:59–71.

Lorenz, K. 1965. *Evolution and modification of behavior.* University of Chicago Press, Chicago.

MacDonald, K.B., and B.E. Ginsburg. 1981. Induction of normal prepubertal behavior in wolves with restricted rearing. *Behavioral and neural biology* 33:133–162.

328 *References*

Macedonia, J.M. 1986. Individuality in a contact call of the ringtailed lemur (*Lemur catta*). *American Journal of Primatology* 11:163–179.

————. 1987. Effects of housing differences upon activity budgets in captive sifakas (*Propithecus verreauxi*). *Zoo Biology* 6:55–67.

Macedonia, J.M., and J.F. Polak. 1989. Visual assessment of avian threat in semi-captive ring tailed lemurs (*Lemur catta*). *Behaviour* 111:291–304.

Macedonia, J.M., and L.L. Taylor. 1985. Subspecific divergence in a loud call of the ruffed lemur (*Varecla variegata*). *American Journal of Primatology* 9:295–304.

Mackie, R.J., K.L. Hamlin, and D.F. Pac. 1982. Mule deer (*Odocileus hemionus*). Pages 862–877 in J.A. Chapman and G.A. Feldhamer, eds., *Wild animals of North America: Biology, management and economics*. Johns Hopkins University Press, Baltimore, Maryland.

Mager, W., and T. Griede. 1986. Using outside areas for tropical primates in the northern hemisphere. Pages 471–477 in K. Benirschke, ed., *Primates: The road to self-sustaining populations*. Springer-Verlag, New York.

Malenky, R.K. 1990. Ecological factors affecting food choice and social organization in Pan paniscus. Ph.D. Dissertation, State University of New York, Stony Brook (*Dissertation Abstracts International* 51(4)B:1613).

Mallis, A. 1982. *Handbook of pest control: The behavior, life history, and control of household pests.* Franzak & Foster, Cleveland.

Mann, M.D., D.A. Crouse, and E.D. Prentice. 1991. Appropriate animal numbers in biomedical research in light of animal welfare considerations. *Laboratory Animal Science* 41:6–14.

Maple, T.L., and T.W. Finlay. 1989. Applied primatology in the modern zoo. *Zoo Biology, Supplement* 1:101–116.

Markowitz, H. 1979. Environmental enrichment and behavioral engineering for captive primates. Pages 217–238 in J. Erwin, T. Maple, and G. Mitchell, eds., *Captivity and behavior: Primates in breeding colonies, laboratories and zoos*. Van Nostrand Reinhold, New York.

————. 1982. *Behavioral enrichment in the zoo.* Van Nostrand Reinhold, New York.

Markowitz, H.., and V.J. Stevens, eds. 1978. *Behavior of captive wild animals.* Nelson Hall, Chicago.

Markowitz, H., and G. Woodworth. 1977. Experimental analysis and control of group behavior. Pages 107–131 in H. Markowitz and V. Stevens, eds. *The behavior of captive wild animals.* Nelson Hall, Chicago.

Marler, P. 1976. Sensory templates in species-specific behavior. Pages 314–329 in J.C. Fentress, ed., *Simpler networks and behavior.* Sinauer Associates, Sunderland, Massachusetts.

Marler, P.R., and W.J. Hamilton. 1966. *Mechanisms of animal behavior.* Wiley, New York.

Marte, E., and F. Halberg. 1961. Circadian susceptibility rhythm of mice to librium. *Federation Proceedings* 20:305 (abstract).

Martin, P., and P. Bateson. 1986. *Measuring behavior.* Cambridge University Press, Cambridge.

Martin, R.D. ed. 1975. *Breeding endangered species in captivity.* Academic Press, London.

Marwine, A., and G. Collier. 1979. The rat at the waterhole. *Journal of Comparative and Physiological Psychology* 93:391–402.

Mason, R.T., R.F. Hoyt, Jr., L.K. Pannell, E.F. Wellner, and B. Demeter. 1991. Cage design and configuration for arboreal reptiles. *Laboratory Animal Science* 41:84–86.

Mason, W.A. 1974. Comparative studies of social behavior in *Callicebus and Saimiri*: behavior of male-female pairs. *Folia Primatologica* 22:1–8.

Mason, W.A., and D.F. Lott. 1976. Ethology and comparative psychology. *Annual Review of Psychology* 27:129–154.

Maynard Smith, J., R. Burian, S. Kauffman, P. Alberch, J. Campbell, B. Goodwin, R. Lande, D. Raup, and L. Wolpert. 1985. Developmental constraints and evolution. *Quarterly Review of Biology* 60:265–287.

Mayr, E. 1974. Behavior programs and evolutionary strategies. *American Scientist* 62:650–659.

McBride, G. 1976. A general theory of social organization and behavior. Pages 86–110 in *Vertebrate social organization. Benchmark Papers in Animal Behavior* 18. Dowden, Hutchinson, and Ross, Stroudsburg, Pennsylvania.

McBride, G., G.W. Arnold, G. Alexander, and J.J. Lynch. 1967. Ecological aspects of the behavior of domestic animals. *Proceedings of the ecological society of Australia* 2:133–165.

McClintock, M.K. 1981. Simplicity from complexity: a naturalistic approach to behavior and neuroendocrine function. Pages 1–19 in I. Silverman, ed., *New directions for methodology of social and behavioral science*, volume 8. Jossey-Bass, San Francisco.

————. 1984. Group mating in the domestic rat as a context for sexual selection: consequences for the analysis of sexual behavior and neuroendocrine responses. *Advances in the Study of Behavior* 14:1–50.

McClintock, M.K., and J.J. Anisko. 1982. Group mating among Norway rats I. Sex differences in the pattern and neuroendocrine consequences of copulation. *Animal Behaviour* 30:398–409 .

McConnell, P. 1991. Lessons from animal trainers: the effect of acoustic structure on an animal's response . Pages 165–187 in P.P.G. Bateson and P.H. Klopfer, eds., *Perspectives in ethology*, vol. 9, *Human understanding and animal awareness*. Plenum Press, New York.

McConnell , P . B ., and C . T . Snowdon. 1986. Vocal interactions between unfamiliar groups of captive cotton-top tamarins. *Behaviour* 97:273–296.

McCord, C.M., and J. E. Cardoza. 1982 . Bobcat and lynx. Pages 728–766 in J.A. Chapman and G.A. Feldhamer, eds., *Wild animals of North America: Biology, management and economics.* Johns Hopkins University Press, Baltimore, Maryland.

McGrew, W.C., J.A. Brennan, and J. Russell. 1986. An artificial "gum-tree" for marmosets (*Callithrix j. jacchus*). *Zoo Biology* 5:45–50.

McGrew, W. C., and E. C. McLuckie. 1986. Pilopatry and dispersion in the cotton-top tamarin, *Saguinus (o.) oedipus*: an attempted laboratory simulation. *International Journal of Primatology* 7:401–422.

McGuire, B., and M. Novak. 1984. A comparison of maternal behavior in the meadow vole (*Microtus pennsylvanicus*), prairie vole (*M. ochrogaster*), and pine vole (*M. pinetorum*). *Animal Behaviour* 32:1132–1141.

McGuire, B., and M. Novak. 1986. Parental care and its relationship to social organization in the montane vole (*Microtus montanus*). *Journal of Mammalogy* 67:305–311.

McIver, A.H. 1965. Reactivity to stress in rats as a function of infantile experience. *Proceedings of the Society for Experimental Biology and Medicine* 119:757–759.

McKenzie, S.M., A.S. Chamove, and A.T.C. Feistner. 1986. Floor coverings and hanging screens alter arboreal monkey behavior. *Zoo Biology* 5:339–348.

McNab, B.K. 1966. An analysis of the body temperature of birds. *Condor* 68:47–55.

McNab, B.K., and P.C. Wright. 1987. Temperature regulation and oxygen consumption in the Phillipine tarsier *Tarsius syrichta*. *Physiological Zoology* 60:596–600.

McNeely, J.A., K.R. Miller, W.V. Reid, R.A. Mittermeier, and T.B. Werner. 1990. *Conserving the world's biological diversity*. IUCN, Gland, Switzerland, World Resources Institute, Conservation International, World Wildlife Fund—U.S., and the World Bank, Washington, D.C.

McSheehy, T. 1983. Overview of the state of the art in environmental monitoring. Pages 161–181 in E.C. Melby, Jr. and M.W. Balk, eds., *The importance of laboratory animal genetics, health, and the environment in biomedical research*. Academic Press, Orlando, Florida.

Medawar, P. 1972. The hope of progress. Methuen, London.

Melampy, R.M., and L.A. Maynard. 1937. Nutrition studies with the cockroach (*Blattella germanica*). *Physiological Zoology* 10:36–44.

Mellgren, R.L., L. Misasi, and S.W. Brown. 1984. Optimal foraging theory: prey density and travel requirements in *Rattus norvegicus*. *Journal of Comparative Psychology* 98:142–153.

Melnick, D., and K. Kidd. 1983. The genetic consequences of social group fission in a wild population of rhesus monkeys (*Macaca mulata*). *Behavioral Ecology and Sociobiology* 12:229–236.

Meltzer, A., G. Goodman, and J. Fistool. 1982. Thermoneutral zone and resting metabolic rate of growing white leghorn-type chickens. *British Poultry Science* 23:383–391.

Menzel, E.W. 1964. Patterns of responsiveness in chimpanzees reared through infancy under conditions of environmental restrictions. *Psychologische Forschung* 27:337–365.

———. 1969. Naturalistic and experimental approaches to primate behavior. Pages 78–121 in E. Willems and H. Rausch, eds., *Naturalistic viewpoints in psychological research*. Holt, Rinehart & Winston, New York.

———. 1971. Communication about the environment in a group of young chimpanzees. *Folia Primatologica* 15:220–232.

———. 1973. Chimpanzee spatial memory organization. *Science* 182:943–945.

————. 1974. A group of young chimpanzees in a one-acre field. Pages 83–153 in A.M. Schrier and F. Stollnitz, eds., *Behavior of nonhuman primates*, vol. 5. Academic Press, New York.

————. 1984. Spatial cognition and memory in captive chimpanzees. Pages 509–531 in P. Marler and H.S. Terrace, eds., *The biology of learning*. Springer-Verlag, Berlin.

Menzel, E.W., and C. Juno. 1982. Marmosets (*Saguinus fuscicollis*): are learning sets learned? *Science* 217:750–752.

Menzel, E.W., and C. Juno. 1985. Social foraging in marmoset monkeys and the question of intelligence. *Philosophical Transactions of the Royal Society of London*, Series B [*Biological Sciences*] 308:145–158.

Menzel, E.W., and C.R. Menzel. 1979. Cognitive, developmental and social aspects of responsiveness to novel objects in a family group of marmosets (*Saguinus fusicollis*). *Behaviour* 70:251–278.

Menzel, E.W., and E.J. Wyers. 1981. Cognitive aspects of foraging behavior. Pages 355–377 in A.C. Kamil and T.D. Sargent, eds., *Foraging behavior: Ecological, ethological, and psychological approaches*. Garland STPM, New York.

Midgley, A.R., Jr., G.D. Niswender, and R.W. Rebar. 1969. Principles for the assessment of radioimmunoassay methods (precision, accuracy, sensitivity, specificity). Pages 163–180 in A. Diczfalusy, ed., *Karolinska symposia on research methods in reproductive endocrinology, 1st symposium*. Karolinska Institutet, Stockholm, Sweden.

Miles, R.C., and D.R. Meyer. 1956. Learning sets in marmosets. *Journal of Comparative and Physiological Psychology* 49:219–222.

Miller, D.B. 1977a. Social displays of mallard ducks (*Anas platyrhynchos*): effects of domestication. *Journal of Comparative and Physiological Psychology* 91:221–232.

————. 1977b. Roles of naturalistic observation in comparative psychology. *American Psychologist* 32:211–219.

Miller, L., and D. Quiatt. 1985. Tool use by a captive orangutan. *Laboratory Primate Newsletter* 24:10.

Miller, S.D. 1980. The ecology of the bobcat in southern Alabama. Ph.D. Dissertation. Auburn University, Auburn, Alabama (*Dissertation Abstracts International* 41(6)B:1997).

Milligan, S.R. 1976. Pregnancy blocking in the vole, *Microtus agrestis* I. Effect of the social environment. *Journal of Reproduction and Fertility* 46:91–95.

Milliken, G.W., J.P. Ward, and C.J. Erickson. 1991. Independent digit control in foraging by the aye-aye (*Daubentonia madagascariensis*). *Folia Primatolgica* 56:219–224.

Milton, K. 1980. *The foraging strategy of howler monkeys: A study in primate economics.* Columbia University Press, New York.

———. 1984. The role of food-processing factors in primate food choice. Pages 249–279 in P.S. Rodman and J.G.H. Cant, eds., *Adaptations for foraging in nonhuman primates: Contributions to an organismal biology of prosimians, monkeys, and apes.* Columbia University Press, New York.

Mineka, S., and M. Cook. 1988. Social learning and the acquisition of snake fear in monkeys. Pages 51–73 in T.R. Zentall and B.G. Galef, Jr., eds., *Social learning. Psychological and biological perspectives.* Lawrence Erlbaum, Hillsdale, New Jersey.

Mineka, S., M. Davidson, M. Cook, and R. Keir. 1984. Observational conditioning of snake fear in rhesus monkeys. *Journal of Abnormal Psychology* 93:355–372.

Mineka, S., M. Gunnar, and M. Champoux. 1986. Control and early socioemotional development: infant rhesus monkeys reared in controllable versus uncontrollable environments. *Child Development* 57:1241–1256.

Mineka, S., and R.W. Henderson. 1985. Controllability and predictability in acquired motivation. *Annual Review of Psychology* 36:495–529.

Mineka, S., and R. Keir. 1983. The effects of flooding on reducing snake fear in rhesus monkeys: six month follow-up and further flooding. *Behavioral Research and Therapy* 21:527–535.

Mineka, S., R. Keir, and V. Price. 1980. Fear of snakes in wild- and lab-reared rhesus monkeys. *Animal Learning and Behavior* 8:653–663.

Missakian, E.A. 1972. Geneological and cross-geneological dominance relation in a group of free-ranging rhesus monkeys (*Macaca mulatta*) on Cayo Santiago. *Primates* 13:169–180.

Mittermeier, R.A., J.F. Oates, A.E. Eudey, and J. Thornback. 1986. Primate Conservation. Pages 3–72 in G. Mitchell and J. Erwin, eds., *Comparative primate biology*, volume 2A: *Behavior, conservation, and ecology.* Alan R. Liss, New York.

Middleton, W.E.K. 1966. *A history of the thermometer: And its use in meteorology.* Johns Hopkins Press, Baltimore, Maryland.

Mizuno, T., and H. Tsuji. 1974. Harbouring behavior of three species of cockroaches, *Periplaneta americana, P. jabonica, and Blattella germanica* [in Japanese, with English summary]. *Japanese Journal of Sanitation Zoology* 24:237–240.

Moberg, G.P., 1985a. *Animal stress.* American Physiological Society, Bethesda, Maryland.

————. 1985b. Biological response to stress: key to assessment of animal well-being? Pages 27–49 in G. Moberg, ed., *Animal stress.* American Physiological Society, Bethesda, Maryland.

————. 1987. A model for assessing the impact of behavioral stress on domestic animals. *Journal of Animal Science* 65:1228–1235.

Molzen, E., and J. French. 1989. The problem of foraging in captive callitrichid primates: behavioral time budgets and foraging skills. Pages 89–101 in E. Segal, ed., *Housing, care, and psychological well-being of captive and laboratory primates.* Noyes Publications, Park Ridge, New Jersey.

Montagna, W. 1976. Diseases common to monkeys and man. Pages 99–128 in W. Montagna, *Nonhuman primates in biomedical research.* University of Minnesota Press, Minneapolis.

Morton, D.B., and P.H.M. Griffiths. 1985. Guidelines on the recognition of pain, distress and discomfort in experimental animals and an hypothesis for assessment. *Veterinary Record* 116:431–436.

Moynihan, M. 1964. Some behavior patterns of Platyrrhine monkeys, I. The night monkey (*Aotus trivirgatus*). *Smithsonian Miscellaneous Collection* 146:1–84.

Munkelt, F.H. 1938. Odor control in animal laboratories. *Heat Piping Air Conditioning* 10:289–291.

————. 1948. Air purification and deodorization by use of activated carbon. *Refrigeration Engineers* 56:222–229.

Murakami, H. 1971. Differences between internal and external environments of the mouse cage. *Laboratory Animal Science* 21:680–684.

Myers, K., C.S. Hale, R. Mykytowycz, and R.L. Hughes. 1971. The effects of varying density and space on sociality and health in animals. Pages 148–187 in A.H. Esser, ed., *Behavior and environment: The use of space by animals and man.* Plenum Press, New York.

Nagel, T. 1974. What is it like to be a bat? *Philosophical Review* 83:435–450.

Nagel, T. 1986. *The view from nowhere.* Oxford University Press, New York.

Nash, L.T., and S.-M. Chilton. 1986. Space or novelty? effects of altered cage size on Galago behavior. *American Journal of Primatology* 10:37–49.

Neimitz, C. 1984. *The biology of tarsiers.* Gustav Fischer, Stuttgart.

Neisser, U. 1976. *Cognition and reality: Principles and implications of cognitive psychology.* Freeman, San Francisco.

Nelson, J.B. 1960. The problems of disease and quality in laboratory animals. *Journal of Medical Education* 35:34–44.

Newberne, P.M., and J.G. Fox. 1978. Chemicals and toxins in the animal facility. Pages 118–138 in E. Besch, Chairman, *Laboratory animal housing.* Institute of Laboratory Animal Resources, National Academy of Sciences, Washington, D.C.

Newman, J.D., and D. Symmes. 1982. Inheritance and experience in the acquisition of primate acoustic behavior. Pages 259–278 in C.T. Snowdon, C.H. Brown, and M.R. Petersen, eds., *Primate communication.* Cambridge University Press, Cambridge, England.

Neyman, P.F. 1977. Aspects of the ecology and social organization of free-ranging cotton-top tamarins (*Saguinus oedipus*) and the conservation status of the species. Pages 39–71 in D.G. Kleiman, ed., *The biology and conservation of the Callitrichidae.* Smithsonian Institution Press, Washington, D.C.

Nicholls, T.J., and P.D. Handson. 1983. Behavioral change associated with chronic lead poisoning in working dogs. *Veterinary Record* 112:607.

Nicholson, R.I., and J.J. McGlone. 1991. Environmental influences on pig tail biting, tail posture and performance. *Journal of Animal Science* 69, *Supplement* 1:233 (abstract).

Njaa, L.R., F. Utne, and O.R. Braekkan. 1957. Effect of relative humidity on rat breeding and ringtail. *Nature* 180:290–291.

Novak, M.A., and K. Bayne, eds. 1991. Special topic overview: monkey behavior and laboratory issues. *Laboratory Animal Science* 41:306–384.

Novak, M.A., and K.H. Drewsen. 1989. Enriching the lives of captive primates: issues and problems. Pages 161–182 in E. Segal, ed., *Housing, care, and psychological well-being of captive and laboratory primates.* Noyes Publications, Park Ridge, New Jersey.

Novak, M.A., H. Munroe, A. Rulf, P. O'Neill, C. Price, and S. Suomi. 1992. Control over the behavioral repertoire of monkeys living in different environments. *American Journal of Primatology* 27:48 (abstract).

Novak, M.A., P. O'Neill, and S.J. Suomi. 1989. Back to the field: social propensities of lab-born rhesus monkeys living in different environments. *American Journal of Primatology* 18:158 (abstract).

Novak, M.A., P. O'Neill, and S.J. Suomi. In press. Adjustments and adaptations to indoor and outdoor environments: continuity and change in young adult rhesus monkeys. *American Journal of Primatology.*

Novak, M.A., and S.J. Suomi. 1988. Psychological well-being of primates in captivity. *American Psychologist* 43:765–773.

Novak, M.A., and S.J. Suomi. 1991. Social interaction in nonhuman primates: an underlying theme for primate research. *Laboratory Animal Science* 41:308–314.

O'Leary, K.D. 1988. Physical aggression between spouses: a social learning perspective. Pages 31–55 in R.L. Morrison, A.S. Bellack and M. Hersen, eds., *Handbook of family violence.* Plenum Press, New York.

O'Neill, P.L. 1988. Developing effective social and environment enrichment strategies for macaques in captive groups. *Laboratory Animal,* May/June:23–36.

O'Neill, P.L., M.A. Novak, and S.J. Suomi. 1991. Normalizing laboratory-reared rhesus macaque behvior with exposure to enriched outdoor enclosures. *Zoo Biology* 10:237–245.

O'Neill, P.L., and C.S. Price. 1991. Customizing and enrichment program: rhesus monkeys. *Lab Animal* 20:29–40.

O'Steen, W.K. 1970. Retinal and optic nerve serotonin and retinal degeneration as influenced by photoperiod. *Experimental Neurology* 27:194–205.

O'Steen, W.K., C.R. Shear, and K.V. Anderson. 1972. Retinal damage after prolonged exposure to visible light: a light and electron microscope study. *American Journal of Anatomy* 134:5–22.

Orlans, F.B., R.C. Simmonds, and W.J. Dodds, eds. 1987. Effective animal care and use committees. *Laboratory Animal Science*, special issue.

Osborne, S.R. 1977. The free food (contrafreeloading) phenomenon. *Animal Learning & Behavior* 5:221–235.

Ostfeld, R.S. 1985. Limiting resources and territoriality in microtine rodents. *Amerlcan Naturalist* 126:1–15.

Outteridge, P.M. 1985. Veterinary immunology. Academic Press, London.

Overmier, J.B., J. Patterson, and R.M. Wielkiewicz. 1980. Environmental contingencies as sources of stress in animals. Pages 1–37 in S. Levine and H. Ursin, eds., *Coping and health*. Plenum Press, New York.

Overmier, J.B., and M.E.P. Seligman. 1967. Effects of inescapable shock upon subsequent escape and avoidance responding. *Journal of Comparative and Physiological Psychology* 63:28–33.

Paquette, D., and J. Prescott. 1988. Use of novel objects to enhance environments of captive chimpanzees. *Zoo Biology* 7:15–23.

Patterson, R.S., and P.G. Koehler. 1985. Sterility: a practical IPM approach for German cockroach (*Blattella germanica*) control. *Proceedings of the 1st Insect Growth Regulator Symposium* 1:48–60.

Paulk, H.H., H. Dienske, and L.G. Ribbens. 1977. Abnormal behavior in relation to cage size in rhesus monkeys. *Journal of Abnormal Psychology* 86:87–92.

Pearl, M.C., and S.R. Schulman. 1983. Techniques for the analysis of social structure in animal societies. Pages 107–146 in J.S. Rosenblatt, R.A. Hinde, C. Beer, and M.-C. Busnel, eds., *Advances in the Study of Behavior* vol. 13. Academic Press, New York.

Pereira, M.E. 1991. Asynchrony within estrous synchrony in ringtailed lemurs (Primates: Lemuridae). *Physiology and Behavior* 49:47–52.

Pereira, M.E., and J. Altmann. 1985. Development of social behavior of free-living non-human primates. Pages 217–309 in E.S. Watts, ed., *Nonhuman primate models for human growth and development*. Alan R. Liss, New York.

Pereira, M.E., and S.A. Altmann. 1983. *Statement for Institute of Laboratory Animal Resources' public meeting*: 12 July, Rosemont, Illinois. (Copies available upon request to senior author.)

Pereira, M.E., and M.K. Izard. 1989. Lactation and care for unrelated infants in forest-living ringtailed lemurs. *American Journal of Primatology* 18:101–108.

Pereira, M.E., R. Kaufman, P.M. Kappeler, and D.J. Overdorff. 1990. Female dominance does not characterize all of the Lemuridae. *Folia Primatologica* 55:96–103.

Pereira, M.E., A. Klepper, and E.L. Simons. 1987. Tactics of care for young infants by forest-living ruffed lemurs (*Varecia variegata variegata*): ground nests, parking, and biparental guarding. *American Journal of Primatology* 13:129–144.

Pereira, M.E., J.M. Macedonia, D.M. Haring, and E.L. Simons. 1989. Maintenance of primates in captivity for research: the need for naturalistic environments. Pages 40-60 in E. Segal, ed., *Housing, care, and psychological well-being of captive and laboratory primates.* Noyes Publications, New York.

Pereira, M.E., M.L. Seeligson, and J.M. Macedonia. 1988. The behavioral repertoire of the black-and-white ruffed lemur, *Varecia variegata variegata* (Primates: Lemuridae). *Folia Primatologica* 51:1–32.

Pereira, M.E., and M.L. Weiss. 1991. Female mate choice, male migration, and the threat of infanticide in ringtailed lemurs. *Behavioral Ecology and Sociobiology* 28:141–152.

Peres, C. 1989. Costs and benefits of territorial defense in wild golden lion tamarins, *Leontopithecus rosalia. Behavioral Ecology and Sociobiology* 25:227–233.

Peres, C., J. Dietz, and L. Pinder. In press. Habitat preference, foraging and feeding, and time budget for golden lion tamarins: daily activity cycles of free-ranging groups. In D. Kleiman, ed., *A Case Study in conservation biology: The golden lion tamarin.* Smithsonian Institution Press, Washington, D.C.

Perry, J.M., M.K. Izard, and P.A. Fail. 1992. As assessment of reproductive competence in captive Lemur mongoz (*mongoose lemur*). *Zoo Biology* 11:81–97.

Perry-Richardson, J.J., and C.S. Ivanyi. 1994. Captive design for reptiles and amphibians. In E.F. Gibbons, Jr., B.S. Durrant and J. Demarest, eds., *Conservation of endangered species in captivity: An interdisciplinary approach.* The State University of New York Press, Albany, New York.

Petersen, M.R., M.D. Beecher, S.R. Zoloth, S. Green, P. Marler, D.B. Moody, and W.C. Stebbins. 1984. Neural lateralization of vocalizations by Japanese macaques: communicative significance is more important than acoustic structure. *Behavioral Neuroscience* 98:779–790.

Petrinovich, L. 1990. Avian song development: methodological and conceptual issues. Pages 340–359 in D.A. Dewsbury, ed., *Contemporary issues in comparative psychology.* Sinauer, Sunderland.

Pevet, P. 1987. Environmental control of the annual reproductive cycle in mammals. Pages 82–100 in P. Pevet, ed., *Comparative physiology of environmental adaptations: Adaptations to climatic changes.* Karger, Basel, Switzerland.

Phillips, G.B., M. Reitman, C.L. Mullican, and G.D. Gardner, Jr. 1957. Applications of germicidal ultraviolet in infectious disease laboratories III. The use of ultraviolet barriers on animal cage racks. *Proceedings of the Animal Care Panel* 7:235–244.

Phillips, M.T., and J.A. Sechzer. 1989. *Animal research and ethical conflict: An analysis of the scientific literature 1966–1986.* Springer-Verlag, New York.

Pierce, G.J., and J.G. Ollason. 1987. Eight reasons why optimal foraging theory is a complete waste of time. *Oikos* 49:111–118.

Pinel, J.P.J., and M.J. Mana. 1989. Adaptive interactions of rats with dangerous inanimate objects: support for a cognitive theory of defensive behavior. Pages 137–150 in R.J. Blanchard, P.F. Brain, D.C. Blanchard and S. Parmigiani, eds., *Ethoexperimental approaches to the study of behavior.* Dordrecht, Kluwer.

Plimpton, E.H., K.B. Swartz, and L.A. Rosenblum. 1981. The effects of foraging demand on social interactions in a laboratory group of bonnet macaques. *International Journal of Primatology* 2:175–185.

Pola, Y.V., and C.T. Snowdon. 1975. The vocalizations of pygmy marmosets (*Cebuella pygmaea*). *Animal Behaviour* 23:826–842.

Pollock, J.I. 1986a. Primates and conservation priorities in Madagascar. *Oryx* 20:209–216.

———. 1986b. The management of prosimians in captivity for conservation and research. Pages 269–288 in K. Benirschke, ed., *Primates: The road to self-sustaining populations.* Springer-Verlag, New York.

Poole, T.B. 1987. Social behavior of a group of orangutans (*Pongo pygmeaus*) on an artificial island in Singapore Zoological Gardens. *Zoo Biology* 6:315–330.

Poole, T.B., and H.D.R. Morgan. 1976. Social and territorial behavior of laboratory mice (*Mus musculus L.*) in small complex areas. *Animal Behaviour* 24:476–480.

Port, C.D., and J.P. Kaltenbach. 1969. The effect of corncob bedding on reproductivity and leucine incorporation in mice. *Laboratory Animal Care* 19:46–49.

Post, D.G. 1984. Is optimization the optimal approach to primate foraging? Pages 280–303 in P.S. Rodman and G.H. Cant, eds., *Adaptations for foraging in nonhuman primates: Contributions to an organismal biology of prosimians, monkeys, and apes.* Columbia University Press, New York.

Post, D.G., G. Hausfater, and S.A. McCuskey. 1980. Feeding behavior of yellow baboons (*Papio cynocephalus*): relationships to age, gender and dominance rank. *Folia Primatologica* 34:170–195.

Potts, W.K., C.J. Manning, and E.K. Wakeland. 1989. Evidence for over-dominance and reproductive selection in the maintenance of MHC polymorphisms in semi-natural populations of *Mus. Journal of Cellular Biochemistry, Supplement* 13C:137 (abstract).

Potts, W.K., C.J. Manning, and E.K. Wakeland. 1991. Mating patterns in seminatural populations of mice influenced by MHC genotype. *Nature* 352:619–621.

Potts, W.K., and E.K. Wakeland. 1990. Evolutionary forces maintaining genetic diversity at the major histocompatibility complex (MHC). *Trends in Ecology and Evolution* 5:181–187.

Pratt, B.L., and B.D. Goldman. 1986. Activity rhythms and photoperiodism of Syrian hamsters in a simulated burrow system. *Physiology and Behavior* 36:83–89.

Price, E.O. 1977. Burrowing in wild and domestic Norway rats. *Journal of Mammology* 58:239–240.

———. 1984. Behavioral aspects of animal domestication. *Quarterly Review of Biology* 59:1–32.

———. 1985. Evolutionary and ontogenetic determinants of animal suffering and well-being. Pages 15–26 in G.P. Moberg, ed., *Animal stress.* American Physiological Society, Bethesda, Maryland.

Prosser, C.L. 1964. Perspectives of adaptation: theoretical aspects. Pages 11–25 in D.B. Dill, E.F. Adolph, and C.G. Wilber, eds., *Handbook of physiology.* Section 4: *Adaptation to the environment.* American Physiological Society, Washington, D.C.

Prost, J.H. 1965. A definitional system for the classification of primate locomotion. *American Anthropology* 67:1198–1214.

Public Health Service. 1986. *Public health service policy on humane care and use of laboratory animals.* Office of Protection from Research Risks, Department of Health and Human Services, Public Health Service, National Institutes of Health, Bethesda, Maryland.

Rai, R.M., A.P. Singh, T.N. Upadhayay, S.K.B. Patil, and H.S. Nayar. 1981. Biochemical effects of chronic exposure to noise in man. *International Archives of Occupational Environmental Health* 48:331–337.

Ralls, K. 1971. Mammalian scent marking. *Science* 171:443–449.

Ramirez, J.M., R.A. Hinde, and J. Groebel, eds. 1987. *Essays on violence.* Publicaciones de la Universidad de Sevilla. Sevilla, Spain.

Ramsey, J.A. 1935. The evaporation of water from the cockroach. *Journal of Experimental Biology* 12:373–383.

Rasmussen, A.F., Jr., J.T. Marsh, and N.Q. Brill. 1957. Increased susceptibility to herpes simplex in mice subjected to avoidance-learning stress or restraint. *Proceedings of the Society for Experimental Biology and Medicine* 96:183–189.

Rasmussen, D.T. 1985. A comparative study of breeding seasonality and litter size in eleven taxa of captive lemurs (*Lemur* and *Varecia*). *International Journal of Primatology* 6:501–517.

Rasmussen, D.T., and M.K. Izard. 1988. Scaling of growth and life history traits relative to body size, brain size, and metabolic rate in *lorises* and *galagos* (Lorisidae, Primates). *American Journal of Physical Anthropology* 75:357–367.

Ratcliffe, H.L. 1968. Environment, behavior, and disease. Pages 161–228 in E. Stellar and J.M. Sprague, eds., *Progress in physiological psychology*, volume 2. Academic Press, New York.

Reed, R.P. 1972. High resolution thermometry in the biological, context—some problems and techniques. Pages 2137–2149 in D.I. Finch, G.W. Burns, R.L. Berger and T.E. Van Zandt, eds., *Temperature: Its measurement and control in science and industry*, volume 4, part 3. *Thermocouples, biology and medicine, and geophysics and space.* Instrument Society of America, Pittsburgh.

Reierson, D.A., and M.K. Rust. 1977. Trapping, flushing, counting German cockroaches. *Pest Control* 45:40, 42–44.

Reiling, G.H. 1989. Light source characteristics and selection. Pages 55–58 in D.C. Holley, C.M. Winget and H.A. Leon, eds., *Lighting require-*

ments in microgravity—rodents and nonhuman primates. National Aeronautics and Space Administration, Ames Research Center, Moffett Field, California.

Reimers, T.J., and S.V. Lamb. 1991. Radioimmunoassay of hormones in laboratory animals for diagnostics & research. *Lab Animal* 20:32–38.

Reinhardt, V., W.D. Houser, D. Cowley, and M. Champoux. 1987. Preliminary comments on environmental enrichment with branches for individually caged rhesus monkeys. *Laboratory Primate Newsletter* 26:1–3.

Reite, M. 1985. Implantable biotelemetry and social separation in monkeys. Pages 141–160 in G.P. Morberg, ed., *Animal stress.* American Physiological Society, Bethesda, Maryland.

Reiter, J., K. Panken, and B. Le Boeuf. 1981. Female competition and reproductive success in northern elephant seals. *Animal Behaviour* 29:670–687.

Remfry, T. 1987. Ethical aspects of animal experimentation. Pages 5–19 in A.A. Tuffery, ed., *Laboratory animals. An introduction for new experimenters.* John Wiley, Chechester, Great Britain.

Renquist, D.M., and R.A. Whitney, Jr. 1987. Zoonoses acquired from pet primates. *Veterinary Clinics of North America: Small Animal Practice* 17:219–240.

Richard, A.F. 1985. *Primates in nature.* W.H. Freeman, New York.

Richard, A.F., S.J. Goldstein, and R.E. Dewar. 1989. Weed macaques: the evolutionary implications of macaque feeding ecology. *International Journal of Primatology* 10:569–594.

Richards, S.A. 1970. The role of hypothalamic temperature in the control of panting in the chicken exposed to heat. *Journal of Physiology* 211:341–358.

Richmond, J.Y. 1991. Responsibilities in animal research. *Lab Animal* 20:41–46.

Riddle, O., G.C. Smith, and F.G. Benedict. 1934. Seasonal and temperature factors and their determination in pigeons of percentage metabolism change per degree of temperature change. *American Journal of Physiology* 107:333–342.

Riess, B.F. 1950. The isolation of factors of learning and native behavior in field and laboratory studies. *Annals of the New York Academy of Science* 50:1093–1103.

————. 1954. The effect of altered environment and of age on mother-young relationships among animals. *Annals of the New York Academy of Science* 57:606–610.

Rijksen, H.D. 1978. A field study on Sumatran orangutans (*Pongo pygmaeus abelii* Lesson 1827): *Ecology, behaviour and conservation*. H. Veenman and Zonen BV, Wageningen, The Netherlands.

Riopelle, A.J., C.W. Hill, and S.-C. Li. 1975. Protein deprivation in primates. V. Fetal mortality and neonatal status of infant monkeys born of deprived mothers. *American Journal of Clinical Nutrition* 28:989–993.

Ritvo, H. 1987. *The animal estate: The english and other creatures in the Victorian Age*. Harvard University Press, Cambridge.

Roberts,M., and B. Cunningham. 1986. Space and substrate use in captive western tarsiers, *Tarsius bancanus*. *International Journal of Primatology* 7:113–130.

Robinson, M. 1988. Building the BioPark. *Zoogoer* 17:4–10.

Rogers, K.J., P.F. Pfolliott, and D.R. Patton. 1978. *Home range and movement of five mule deer in a semidesert grass-shrub community*. U.S. Department of Agriculture Forest Service Research Paper RM-355.

Rohles, F.H., Jr. 1978. The empirical approach to thermal comfort. *American Society for Heating, Refrigerating and Air Conditioning Engineers Transactions* 84:725–732.

Roper, T.J., and E. Polioudakis. 1977. The behaviour of Mongolian gerbils in a semi-natural environment, with special reference to ventral marking, dominance and sociability. *Behaviour* 61:207–237.

Rosenblum, L.A., and I.C. Kaufman. 1967. Laboratory observations of early mother-infant relations in pigtail and bonnet macaques. Pages 33–41 in S.A. Altmann, ed., Social communication among primates. University of Chicago Press, Chicago.

Rosenblum, L.A., H. Kummer, R.D. Nadler, J. Robinson, and S.J. Suomi. 1989. Commentary: interface of field and laboratory-based research in primatology. *American Journal of Primatology* 18:61–64.

Rosenblum, L.A., and J. Smiley. 1984. Therapeutic effects of an imposed foraging task in disturbed monkeys. *Journal of Child Psychology and Psychiatry* 25:485–497.

Rosenzweig, M.R., E.L. Bennett, and M.C. Diamond. 1972. Brain changes in response to experience. *Scientific American* 226:22–29.

Ross, O.B. 1960. Swine housing. *Agricultural Engineering* 41:584–585.

Roth, L.M., and D.W. Alsop. 1978. Toxins of Blattaria. Pages 465–487 in S. Bettini, ed., *Arthropod venoms*. Springer-Verlag, Berlin.

Roth, L.M., and E.R. Willis. 1960. The biotic associations of cockroaches. *Smithsonian Miscellaneous Collections* 141:1–470.

Rothstein, S.I., D.A. Yokel, and R.C. Fleischer. 1988. The agonistic and sexual functions of vocalizations of male brown-headed cowbirds, *Molothrus ater*. *Animal Behavior* 36:73–86.

Roy, M.A., ed. 1980. *Species identity & attachment: A phylogenetic evaluation*. Garland STPM Press, New York.

Runkle, R.S. 1964. Laboratory animal housing II. *Journal of the American Institute of Architects* 41:77–80.

Rusak, B. 1975. Neural control of circadian rhythms in behavior of the golden hamster, *Mesocricetus auratus*. Ph.D. Dissertation, University of California, Berkeley. (*Dissertation Abstracts International* 35(5)B:2565).

Ruskin, R.E., ed. 1965. *Humidity and moisture: Measurement and control in science and industry*, volume one. *Principles and methods of measuring humidity in gases*. Reinhold, New York.

Rust, M.K. 1986. Managing household pests. Pages 335–368 in G.W. Bennett and J.M. Owens, eds., *Advances in urban pest management*. Van Nostrand Reinhold, New York.

Rust, M.K., and D.A. Reierson. 1981. Attraction and performance of insecticidal baits for German cockroach control. *International Pest Control* 23:106–109.

Sachs, B.D., and R.J. Barfield. 1976. Functional analysis of masculine copulatory behavior in the rat. *Advances in the Study of Behavior* 7:91–154.

Sackett, G.P. 1968. Abnormal behavior in laboratory-reared rhesus monkeys. Pages 293–331 in M.W. Fox, ed., *Abnormal behavior in animals*. Saunders, Philadelphia.

Sackett, G.P. ed. 1978. *Observing behavior*, vol. II. *Data collection and analysis methods*. University Park Press, Baltimore.

Sadleir, R.M.F.S. 1975. Role of the environment in the reproduction of mammals in zoos. Pages 151–156 in *Research in zoos and aquariums*. National Academy of Sciences, Washington, D.C.

Samonds, K.W., and D.M. Hegsted. 1978. Protein deficiency and energy restriction in young cebus monkeys. *Proceedings of the National Academy of Sciences, U.S.A.* 75:1600–1604.

Samuel , D. E., and B.B. Nelson. 1982 . Foxes (*Vulpes vulpes*) and allies. Pages 475–490 in J.A. Chapman and G.A. Feldhamer, eds., *Wild animals of North America: Biology, management and economics.* Johns Hopkins University Press, Baltimore, Maryland.

Sanford, J., R. Ewbank, V. Molony, W.D. Tavernor, and O. Uvarov. 1986. Guidelines for the recognition and assessment of pain in animals. *Veterinary Record* 118: 334–338.

Sapoff, M. 1972. Thermistors for biomedical use. Pages 2109–2121 in D.I. Finch, G.W. Burns, R.L. Berger and T.E. Van Zandt, eds., *Temperature: Its measurement and control in science and industry*, volume 4, part 3. *Thermocouples, biology and medicine, and geophysics and space.* Instrument Society of America, Pittsburgh.

Saunders, J.K., Jr. 1963. Movements and activities of the lynx in Newfoundland. *Journal of Wildlife Management* 27:390–400.

Savage, A., L.A. Dronzek, and C.T. Snowdon. 1987. Color discrimination by the cotton-top tamarin (*Saguinus oedipus oedipus*) and its relation to fruit coloration. *Folia Primatologica* 49:57–69.

Savage, A., T.E. Ziegler, and C.T. Snowdon. 1988. Sociosexual development, pair bond formation, and mechanisms of fertility suppression in female cotton-top tamarins (*Saguinus oedipus oedipus*). *American Journal of Primatology* 14:345–359.

Sawrey, D.K., and D.A. Dewsbury. 1981. Effects of space on the copulatory behavior of deer mice (*Peromyscus maniculatus*). *Bulletin of the Psychonomic Society* 17:249–251.

Schal, C., J.-Y. Gautier, and W.J. Bell. 1984. Behavioral ecology of cockroaches. *Biological Reviews* 59:209–254.

Schapiro, S., L. Brent, M. Bloomsmith, and W. Satterfield. 1991. Enrichment devices for nonhuman primates. *Lab Animal* 20:22–28.

Schassburger, R.M. 1987. Design of a wolf woods: planning for research capabilities in zoological gardens. Pages 9–30 in H. Frank, ed., *Man and wolf. Advances, issues, and problems in captive wolf research.* Dr. W. Junk, Dordrecht, The Netherlands.

Schierwater, B., and H. Klingel. 1986. Energy costs of reproduction in the Djungarian hamster *Phodopus sungorus* under laboratory and seminatural conditions. *Oecologia* 69:144–147.

Schilling, A. 1979. Olfactory communication in prosimians. Pages 461–542 in G.A. Doyle and R.D. Martin, eds., *The study of prosimian behavior.* Academic Press, New York.

Schmidt-Nielsen, K., B. Schmidt-Nielsen, S.J. Jarnum, and T.R. Houpt. 1957. Body temperature of the camel and its relation to water economy. *American Journal of Physiology* 188:103–112.

Scholander, P.F., R. Hock, V. Walters, F. Johnson, and L. Irving. 1950. Heat regulation in some arctic and tropical mammals and birds. *Biological Bulletin* 99:237–258.

Scott, H.G. 1991. Design and construction: building-out pests. Pages 331–334 in J.R. Gorham, ed. *Ecology and management of food-industry pests.* FDA Technical Bulletin 4, Association of Official Analytical Chemists, Arlington, Virginia.

Selous, F.C. 1967. *A Hunter's Wanderings in Africa.* Arno Press, New York.

Selye, H. 1936. A syndrome produced by diverse nocuous agents. *Nature* 138:32.

Serrano, L.J. 1971. Carbon dioxide and ammonia in mouse cages: effect of cage covers, population, and activity. *Laboratory Animal Science* 21:75–85.

Setchell, K.D.R., S.J. Gosselin, M.B. Welch, J.O. Johnston, W.F. Balistreri, L.W. Kramer, B.L. Dresser, and M.J. Tarr. 1987. Dietary estrogens—a probable cause of infertility and liver disease in captive cheetahs. *Gastroenterology* 93:225–233.

Seyfarth, R.M., and D.L. Cheney. 1986. Vocal development in vervet monkeys. *Animal Behaviour* 34:1640–1658.

Shapiro, L.E., and D.A. Dewsbury. 1990. Differences in affiliative behavior, pair bonding, and vaginal cytology in two species of vole (*Microtus ochrogaster* and *M. montanus*). *Journal of Comparative Psychology* 104:268–274.

Sharon, I. M., R. P. Feller, and S. W. Burney. 1971. The effects of lights of different spectra on caries incidence in the golden hamster. *Archives of Oral Biology* 16:1427–1432

Shemano, I. and M. Nickerson. 1958. Effect of ambient temperature on thermal responses to drugs. *Canadian Journal of Biochemistry and Physiology* 36:1243–1249.

Shepard, P. 1978. *Thinking animals: Animals and the development of human intelligence.* Viking Press, New York.

Sherry, D.F. 1985. Food storage by birds and mammals. *Advances in the Study of Behavior* 15:153–188.

Shettleworth, S.J. 1984. Learning and behavioral ecology. Pages 170–194 in J. Krebs and N. Davies, eds., *Behavioral ecology*, 2nd ed. Sinauer, Sunderland, Massachusetts.

———. 1989. Animals foraging in the lab: problems and promises. *Journal of Experimental Psychology: Animal Behavior Processes* 15:81–87.

Shettleworth, S.J., J.R. Krebs, S.D. Healy, and C.M. Thomas. 1990. Spatial memory of food-storing tits (*Parus ater* and *P. atricapillus*): comparison of storing and nonstoring tasks. *Journal of Comparative Psychology* 104:71–81.

Silverman, L., J.W. Whittenberger, and J. Muller. 1949. Physiological response of man to ammonia in low concentrations. *Journal of Industrial Hygiene and Toxicology* 31:74–78

Simpson, G.G. 1962. The status of the study of organisms. *American Scientist* 50:36–45.

Singer, P. 1975. Animal Liberation, cited in *Newsweek*, December 26, 1988, pp. 53–54.

Skinner, B.F. 1938. *The behavior of organisms: An experimental analysis.* Appleton-Century-Crofts, New York.

Slater, P.J.B. 1978. Data collection. Pages 7–24 in P.W. Colgan, ed., *Quantitative ethology.* John Wiley, New York.

Small, W.S. 1900a. an experimental study of the mental processes of the rat. *American Journal of Psychology* 11:133–165.

———. 1900b. Experimental study of the mental processes of the rat II. *American Journal of Psychology* 12:206–239.

Smith, A., Lindburg, D. G., and S. Vehrencamp. 1989. Effect of food preparation on feeding behavior of lion-tailed macaques. *Zoo Biology* 8:57–65.

Smith, E.N. 1980. Physiological radio telemetry of vertebrates. Pages 45–55 in C.J. Amlaner and D.W. MacDonald, eds., *A handbook on biotelemetry and radio tracking.* Pergamon Press, Oxford.

Smith, R.L. 1974. *Ecology and field biology*, 2nd ed. Harper and Row, New York.

Smuts, B.B., D.L. Cheney, R.M. Seyfarth, R.W. Wrangham, and T.T. Struhsaker, eds. 1986. *Primate societies.* University of Chicago Press, Chicago.

Snowdon, C.T. 1987. A naturalistic view of categorical perception. Pages 332–354 in S. Harnad, ed., *Categorical perception: The ground work of cognition.* Cambridge University Press, Cambridge.

————. 1989. The criteria for succesful captive propagation of endangered primates. *Zoo Biology, Supplement* 1:149–161.

Snowdon, C.T., and Y.V. Pola. 1978. Interspecific and intraspecific responses to synthesized pygmy marmoset vocalizations. *Animal Behaviour* 26:192–206.

Snowdon, C.T and A. Savage. 1989. Psychological well-being of captive primates: general considerations and examples from callitrichids. Pages 75–88 in E. Segal, ed., *Housing, care, and psychological well-being of captive and laboratory primates.* Noyes Publications, Park Ridge, New Jersey.

Snowdon, C.T., A. Savage, and P.B. McConnell. 1985. A breeding colony of cotton-top tamarins (*Saguinus oedipus*). *Laboratory Animal Science* 35:477–480.

Snyder, R.L. 1975. Behavioral stress in captive animals. Pages 41–76 in *Research in zoos and aquariums. A symposium.* National Academy of Sciences, Washington, D.C.

Soave, O.A. 1981. Viral infections common to human and nonhuman primates. *Journal of the American Veterinary Medical Association* 179:1385–1388.

Sojka, N.J. 1986. Listen to the animals. *Lab Animal* 15:30–31.

Sokal, R.R., and F.J. Rohlf. 1981. *Biometry,* 2nd. edition. W.H. Freeman, San Francisco.

Soma, L.R. 1987. Assessment of animal pain in experimental animals. *Laboratory Animal Science* 37:71–74.

Southwick, C. H. 1967. An experimental study of intragroup agonistic behavior in rhesus monkeys (*Macaca mulatta*). *Behaviour* 28:182–209.

————. 1969. Aggressive behavior of rhesus monkeys in natural and captive groups. Pages 32–43 in S. Garattini and E.B. Sigg, eds., *Aggressive Behavior.* Excepta Medica Foundation, Amsterdam, and John Wiley, New York.

Southwick, C. H., M.A. Beg, and M.R. Siddiqi. 1965. Rhesus monkeys in North India. Pages 111–159 in I. Devore, ed., *Primate behavior: Field studies of monkeys and apes.* Holt, Rinehart & Winston, New York.

Southwick, C. H., M.F. Siddiqi, M.Y. Farooqui, and B.C. Pal. 1974. Xenophobia among free-ranging rhesus groups in India. Pages 185–209 in R. L. Holloway, ed., *Primate aggression, territoriality, and xenophobia.* Academic Press, New York.

Speller, S.W. 1972. Food ecology and hunting behavior of Denning Arctic Foxes at Aberdeen Lake, Northwest Territories. Ph.D. Dissertation. University of Saskatchewan, Saskatchewan, Saskatoon. (*Dissertation Abstracts International* 33 [3]B:1088).

Spencer-Booth, Y. 1970. The relationship between mammalian young and conspecifics other than mothers and peers: a review. *Advances in the Study of Behavior* 3:120–194.

Spielvogel, L.G. 1978. Opportunities for energy conservation in animal laboratories. Pages 173–78 in E. Besch, Chairman, *Laboratory animal housing.* National Academy of Sciences, Institute of Laboratory Animal Resources, Washington, D.C.

Staal, G.B. 1985. Insect growth regulators in historical perspective. *Proceedings of the 1st Insect Growth Regulator Symposium* 1:1–6.

Stainer, M.W., L.E. Mount, and J. Bligh. 1984. *Energy balance and temperature regulation.* Cambridge University Press, Cambridge.

Stevens, S.S. 1946. On the theory of scales of measurement. *Science* 103:677–680.

———. 1951. Mathematics, measurement, and psychophysics. Pages 1–49 in S.S. Stevens, ed., *Handbook of experimental psychology.* John Wiley, New York.

Stewart, W., R. Crawford, M. Cook, and A. Matchet. 1985. *Animal welfare manual.* U.S. Department of Agriculture-APHIS-Veterinary Service, Bethesda, Maryland.

Stimson, A. 1974. *Photometry and radiometry for engineers.* John Wiley, New York.

Stitt, J.T., and J.D. Hardy. 1971. Thermoregulation in the squirrel monkey (*Saimiri sciureus*). *Journal of Applied Physiology* 31:48–54.

Stohr, W. 1988. Longterm heartrate telemetry in small mammals: a comprehensive approach as a prerequisite for valid results. *Physiology and Behavior* 43:567–576.

Stokols, D. 1972. On the distinction between density and crowding: some implications for future research. *Psychological Review* 79:275–277.

Stoskopf, M.K. 1983. The physiological effects of psychological stress. *Zoo Biology* 2:179–190.

———. 1992. Environmental requirements of freshwater fishes. Pages 545–553 in M.K. Stoskopf, ed., *Fish Medicine*. W.B. Saunders, Philadelphia.

———. 1994. Design of captive environments for fish. In E. Gibbons, B.S. Durrant and J. Demarest, eds., *Conservation of endangered species in captivity: An interdisciplinary approach*. State University of New York Press, Albany, New York.

Stuart, M. D. 1981. Primate locomotion patterns and captive caging. *American Association of Zoological Parks and Aquariums Conference Proceedings*, pp. 13–18.

Stuhlman, R.A., and J.E. Wagner. 1971. Ringtail in *Mystromys albicaudatus:* A case report. *Laboratory Animal Science* 21:585–587.

Sulkin, S.E., and R.M. Pike. 1951. Survey of laboratory-acquired infections. *American Journal of Public Health* 41(7):769–781.

Suomi, S.J. 1976. Mechanisms underlying social development: a reexamination of mother-infant interactions in monkeys. Pages 201–228 in A.D. Pick, ed., *Minnesota symposium on child psychology*, volume 10. University of Minnesota Press, Minneapolis.

Suomi, S.J., and H.F. Harlow. 1972. Social rehabilitation of isolate-reared monkeys. *Developmental Psychology* 6:487–496.

Suomi, S. J., and P. O'Neill. 1987. Longitudinal studies of laboratory-born rhesus monkeys (*Macaca mulatta*) raised in multi-acre outdoor enclosures. *International Journal of Primatology* 8:450 (abstract).

Swan, H. 1974. *Thermoregulation and bioenergetics: Patterns for vertebrate survival.* American Elsevier, New York.

Swartz, K.B., and L.A. Rosenblum. 1981. The social context of parental behavior: a perspective on primate socialization. Pages 417–454 in D.J. Gubernick and P.H. Klopfer, eds., *Parental behavior in mammals*. Plenum Press, New York.

Symington, M. McFarland. 1988. Food competition and foraging party size in the black spider monkey (*Ateles paniscus chamek*). *Behaviour* 105:117–134.

Tanimoto, K. 1943. Ecological studies on plague-carrying animals in Manchuria. *Dobutsugaku Zasshi* 55:111–127 (translation made available by Tumblebrook Farm, West Brookfield, Massachusetts).

Tardif, S.D., C.B. Richter, and R.L. Carson. 1984a. Effects of sibling-rearing experience on future reproductive success in two species of Callitrichidae. *American Journal of Primatology* 6:377–380.

Tardif, S.D., C.B. Richter, and R.L. Carson. 1984b. Reproductive performance of three species of Callitrichidae. *Laboratory Animal Science* 34:272–275.

Tattersall, I. 1982. *The primates of Madagascar.* Columbia University Press, New York.

Taussky, A.H. 1954. A microclorimetric determination of creatinine in the urine by the Jaffe reaction. *Journal of Biological Chemistry* 208:853–861.

Taylor, L., and R.W. Sussman. 1985. A preliminary study of kinship and social organization in a semi-free-ranging group of *Lemur catta*. *International Journal of Primatology* 6:601–614.

Taylor, S.A., and D.A. Dewsbury. 1988. Effects of experience and available cues on estrous versus diestrous preferences in male prairie voles, *Microtus ochrogaster*. Physiology & Behavior 42:379–388.

Taylor, S.A., and J.W. Kijne. 1965. Evaluating thermodynamic properties of soil water. Pages 335–342 in A. Wexler and W.A. Wildhuck, eds., *Humidity and moisture: Measurement and control in science and industry*, volume three. *Fundamentals and standards.* Reinhold, New York.

Thiessen, D.D., and K.O. Maxwell. 1979. A glass rodent enclosure: gerbil city. *Behavior Research Methods Instrumentation* 11:535–537.

Thomas, J.A., and E.C. Birney. 1979. Parental care and mating system of the prairie vole, *Microtus ochrogaster*. *Behavioral Ecology and Sociobiology* 5:171–186.

Thompson, W.R., and L.M. Solomon. 1954. Spontaneous pattern discrimination in the rat. *Journal of Comparative and Physiological Psychology* 47:104–107.

Timberlake, W. 1983. The functional organization of appetitive behavior: Behavior systems and learning. Pages 177–221 in M.D. Zeiler and P.

Harzem, eds., *Advances in analysis of behavior*, volume 3, *Biological factors in learning*. John Wiley, Chichester.

Tinbergen, N. 1951. *The study of instinct*. Clarendon, Oxford.

—————. 1963. On aims and methods of ethology. *Zeitschrift für Tierpsychologie* 20:410–433.

—————. 1968. On war and peace in animals and man. *Science* 160:1411–1418.

Tobach, E., and H. Bloch. 1956. Effect of stress by crowding prior to and following tuberculous infection. *American Journal of Physiology* 187: 399–402 .

Tolman, E.C. 1948. cognitive maps in rats and men. *Psychological Review* 55:189–208.

Tolman, E.C., and E. Brunswik. 1935. The organism and the causal texture of the environment. *Psychological Review* 42:43–77.

Toth, L.A., and G.A. Olson. 1991. Strategies for minimizing pain and distress in laboratory animals. *Lab Animal* 20:33–39.

Tripp, J. K. 1985. Increasing activity in captive orangutans: provision of manipulable and edible materials. *Zoo Biology* 4:225–234.

Trivers, R.L. 1971. The evolution of reciprocal altruism. *Quarterly Review of Biology* 46:35–57.

Tugwell, M. 1987. Analyzing conflicts between the institution, the scientist, the animal care committee, and the concerned public. *Laboratory Animal Science, Special Issue*: 145–147.

Tunnell, G.B. 1977. Three dimensions of naturalness: an expanded definition of field research. *Psychological Bulletin* 84:426–437.

Ucko, P.J., and G.W. Dimbleby, eds. 1969. *The domestication and exploitation of plants and animals*. Aldine, Chicago.

United States Department of Agriculture. 1991. Final Regulations for 1985 Amendments to the Animal Welfare Act, Part 3 Standards. *Federal Register*, February, 15.

United States Public Health Service. 1985. *Guide for the care and use of laboratory animals*. National Institutes of Health Publication 86–23, Bethesda, Maryland.

Van Houten, R., S.L. Weiss, and K.D. O'Leary. 1970. The summation of conditioned suppression. *Journal of the Experimental Analysis of Behavior* 13:75–81.

Van Putten, G. 1969. An investigation into tail-biting among fattening pigs. *British Veterinary Journal* 125:511–516.

Van Winkle, T. J., and M. W. Balk. 1986. Spontaneous corneal opacities in laboratory mice. *Laboratory Animal Science* 36: 248–255 .

Varosi, S.M., R.L. Brigmon, and E.L. Besch. 1990. A simplified telemetry system for monitoring body temperature in small animals. *Laboratory Animal Science* 40: 299–302 .

Vehrencamp, S.L., and J.W. Bradbury. 1984. Mating systems and ecology. Pages 251–278 in J.R. Krebs and N.B. Davies, eds., *Behavioural ecology: An evolutionary approach*. Sinauer Associates, Sunderland, Massachusetts.

Vesell, E. S. 1967. Induction of drug-metabolizing enzymes in liver microsomes of mice and rats by softwood bedding. *Science* 157:1057–1058.

Vesell, E.S., C.M. Lang, W.J. White, G.T. Passananti, R.N. Hill, T.L. Clemens, D.K. Liu,and W.D. Johnson. 1976. Environmental and genetic factors affecting the response of laboratory animals to drugs. *Federation Proceedings* 35:1125–1132.

Vick, L.G. and M.E. Pereira. 1989. Episodic targeting aggression and the histories of *Lemur* social groups. *Behavioral Ecology and Sociobiology* 25:3–12.

Visalberghi, E. 1986. Acquisition of nut-cracking behavior by two capuchin monkeys (*Cebus apella*). *Primate Report* 14:226–227.

————. 1988. Responsiveness to objects in two social groups of tufted capuchin monkeys (*Cebus apella*). *American Journal of Primatology* 15:349–360.

Visalberghi, E., and F. Antinucci. 1986. Tool use in the exploitation of food resources in *Cebus apella*. Pages 57–62 in J.G. Else and P.C. Lee, eds. *Primate ecology and conservation*, volume 2. Cambridge University Press, Cambridge.

Visalberghi, E., and A.F. Vitale. 1990. Coated nuts as an enrichment device to elicit tool use in tufted capuchins (*Cebus apella*). *Zoo Biology* 9:65–72.

Voller, A., and D.E. Bidwell. 1988. Immunoassays with special reference to ELISA methods. Pages 151–158 in G.R. Smith and J.P. Hearn, eds., *Reproduction and disease in captive and wild animals. Symposia of the Zoological Society of London*, number 60. Clarendon Press, Oxford.

Vyas, G., D.P. Stites, and G. Brecher, eds. 1975. *Laboratory diagnosis of immunologic disorders.* Grune & Stratton, New York.

Waage, J.K. 1979. Dual function of the damselfly penis: sperm removal and transfer. *Science* 203:916–918.

Waal, F.B.M. de. 1984. Coping with social tension: sex differences in the effect of food provision to small rhesus monkey groups. *Animal Behaviour* 32:765–773.

Waal, F.B.M. de., and L.M. Luttrell. 1989. Toward a comparative socioecology of the genus *Macaca*: Different dominance styles in rhesus and stumptail monkeys. *American Journal of Primatology* 19:83–109.

Waal, F.B.M. de., J.A.R.A.M. van Hooff, and W.J. Netto. 1976. An ethological analysis of types of agonistic interactions in a captive group of Java-monkeys (*Macaca fascicularis*). *Primates* 17:257–290.

Waal, F.B.M. de., and D. Yoshihara. 1983. Reconciliation and redirected affection in rhesus monkeys. *Behaviour* 85:224–241.

Walcott, T.G. 1980. Heart rate telemetry using micropower integrated circuits. Pages 279–286 in C.J. Amlaner and D.W. MacDonald, eds., *A handbook on biotelemetry and radio tracking.* Pergamon Press, Oxford.

Wallace, M.E. 1982. Some thoughts on the laboratory cage design process. *International Journal for the Study of Animal Problems* 3:234–242.

Wallach, J.D. 1970. Disease problems in group and zoogeographic displays. *Proceedings of the American Association of Zoo Veterinarians*, pp. 12–14. St. Louis.

————. 1976. Gauntlet of the cage. *Proceedings of the American Association of Zoo Veterinarians*, pp. 235–243. St. Louis.

Wallen, K. 1977. Social organization in the dusky-footed woodrat (*Neotoma fuscipes*): Field studies and laboratory experiments. Ph.D. Dissertation, University of California, Berkeley. (*Dissertation Abstracts International* 38(8)B:3951.)

————. 1982. Social organization in the dusky-footed woodrat (*Neotoma fuscipes*): A field and laboratory study. *Animal Behaviour* 30:1171–1182.

Warnke, G.F. 1972. Commercial pyrometers. Pages 503–517 in H. Preston-Thomas, T.P. Murray and R.L. Shepard, eds., *Temperature: Its measurement and control in science and industry*, volume 4, part 1.

Basic methods, scales and fixed points, radiation. Instrument Society of America, Pittsburg.

Wasserman, M., D. Wasserman, Z. Gershon, and L. Zellermayer. 1969. Effects of organochlorine insecticides on body defense systems. *Annals of the New York Academy of Sciences* 160: 393–401.

Webster, A.J.F. 1983. Environmental stress and the physiology, performance and health of ruminants. *Journal of Animal Science* 57:1584–1593.

Wecker, S.C. 1963. The role of experience in habitat selection by the prairie deer mouse, *Peromyscus maniculatus bairdi. Ecological Monographs* 33:307–325.

Wecker, S.C. 1964. Habitat selection. *Scientific American* 211:109–116.

Weichbrod, R.H., C.F. Cisor, J.G. Miller, R.C. Simmonds, A.P. Alvares, and T.H. Ueng. 1988. Effects of cage beddings on microsomal oxidative enzymes in rat liver. *Laboratory Animal Science* 38:296–298.

Weihe, W.H. 1971. The significance of the physical environment for the health and state of adaptation of laboratory animals. Pages 353–178 in *Defining the laboratory animal.* National Academy of Sciences, Washington, D.C.

———. 1976a. The effect of light on animals. Pages 63–76 in T. McSheehy, ed., *Control of the animal house environment. Laboratory animal handbooks 7.* Laboratory Animals, London.

———. 1976a. The effects on animals of changes in ambient temperature and humidity. Pages 41–50 in T. McSheehy, ed., *Laboratory Animal Handbooks 7. Control of the Animal House Environment.* Laboratory Animals, London.

———. 1988. Housing conditions and experimental results. Pages 245–254 in A.C. Beynen and H.A. Solleveld, eds., *New developments in biosciences: Their implications for laboratory animal science.* Martinus Nijhoff, Dordrecht, the Netherlands.

Welles, R.E., and F.B. Welles. 1961. The Bighorn of Death Valley. *U.S. National Park Service Fauna Series 6.*

Werner, F. 1984. Non-hemoparasitic protozoans. Pages 306–324 in G.L. Hoff, F.L. Frye and E.R. Jacobson, eds., *Diseases of amphibians and reptiles.* Plenum Press, New York.

Werner, L.L. 1983. Immunologic diseases affecting internal organ systems. Page 2158–2185 in S.J. Ettinger, ed., *Textbook of Veterinary Internal Medicine.* W.B. Saunders, Philadelphia.

West, M.J., and A.P. King. 1980. Enriching cowbird song by social deprivation. *Journal of Comparative and Physiological Psychology* 94:263–270.

West, M.J., and A.P. King. 1985. Social guidance of vocal learning by female cowbirds: validating its functional significance. *Zeitschrift für Tierpsychologie* 70:225–235.

West, M.J., and A.P. King. 1986. Song repertoire development in male cowbirds (*Molothrus ater*): its relationship to female assessment of song potency. *Journal of Comparative Psychology* 100:296–303.

West, M.J., and A.P. King. 1988a. Vocalizations of juvenile cowbirds (*Molothrus ater ater*) evoke copulatory responses from females. *Developmental Psychobiology* 21:543–552.

West, M.J., and A.P. King. 1988b. Female visual displays affect the development of male song in the cowbird. *Nature* 334:244–246.

West, M.J., A.P. King, and D.H. Eastzer. 1981. Validating the female bioassay of cowbird song: relating differences in song potency to mating success. *Animal Behaviour* 29:490–501.

West, M.J., A.P. King, D.H. Eastzer, and J.E.R. Staddon. 1979. Bioassay of isolate cowbird song. *Journal of Comparative and Physiological Psychology* 93:124–133.

West, M.J., A.P. King, and T.H. Harrocks. 1983. Cultural transmission of cowbird song (*Molothrus ater*): measuring its development and outcome. *Journal of Comparatrive Psychology* 97:327–337.

West, M.J., A.N. Stroud, and A.P. King. 1983. Mimicry of the human voice by European starlings: the role of social interaction. *Wilson Bulletin* 95:635–640.

Westergaard, G. C., and D.M. Fragaszy. 1985. Effects of manipulatable objects on the activity of captive capuchin monkeys (*Cebus apella*). *Zoo Biology* 4:317–327.

Westergaard, G. C., and D.M. Fragaszy. 1987. The manufacture and use of tools by capuchin monkeys (*Cebus apella*). *Journal of Comparative Psychology* 101:159–168.

Westergaard, G. C., and T. Lindquist. 1987. Manipulation of objects in a captive group of lion-tailed macaques (*Macacas silenus*). *American Journal of Primatology* 12:231–234.

Western, D., and M.C. Pearl, eds. 1989. *Conservation for the twenty-first century*. Oxford University Press, New York.

Westneat, D.F. 1990. Genetic parentage in the Indigo banting: a study using DNA fingerprinting. *Behavioral Ecology and Sociobiology* 27:67–76.

Wexler, A.S.B. 1970. Measurement of humidity in the free atmosphere near the surface of the earth. Pages 262–282 in *Meteorological observations and instrumentation.* Meteorological Monograph 33. American Meteorological Society, Boston, Masachusetts.

White, F.J. 1991. Social organization, feeding ecology,and reproductive strategy of ruffed lemurs, *Varecia variegata.* Pages 81–84 in A. Ehara, T. Kimura, O. Takenaka, and M. Iwamoto, eds., *Primatology today. Proceedings of the XIII Congress of the international primatological society, Nagoya and Ryoto, 18–24 July, 1990.* Elsevier Science, Amsterdam, The Netherlands.

White, G.C., and R.A. Garrott. 1990. *Analysis of wildlife radiotracking data.* Academic Press, San Diego.

White, N.R., and R.J. Barfield. 1987. Role of the ultrasonic vocalization of the female rat (*Rattus norvegicus*) in sexual behavior. *Journal of Comparative Psychology* 101:73–81.

White, W.J. 1982. Energy savings in the animal facility opportunities & limitations. *Lab Animal* 11:28–33.

Whitten, W.K., and F.H. Bronson. 1970. The role of pheromones in mammalian reproduction. Pages 309–325 in J.W. Johnston, Jr., D.G. Moulton, and A. Turk, eds., *Advances in chemoreception.* vol. 1. *Communication by chemical signals.* Appleton-Century-Crofts, New York.

Widowski, T.M., T.E. Ziegler, A.M. Elowson, and C.T. Snowdon. 1990. The role of males in the stimulation of reproductive function in female cotton-top tamarins, *Saguinus o. oedipus. Animal Behaviour* 40:731–741.

Wilber, C.G. 1964. Animals in aquatic environments: Introduction. Pages 661–668 in D.B. Dill, E.F. Adolph, and C.G. Wilber, eds., *Handbook of Physiology. Section 4: Adaptation to the environment.* American Physiological Society, Washington, D.C.

Wille, J. 1920. Biologie und bekämpfung der deutschen schabe (*Phyllodromia germanica* L.). *Monog zur angew, Ent Beihefte, I. sur Zeits. f. angew. Ent.,* 7:1–140.

Willis, E.R. and N. Lewis. 1957. The longevity of starved cockroaches. *Journal of Economic Entomology* 50:438–440.

Wilson, E.O., ed. 1989. *Biodiversity*. National Academy Press,Washington, D.C.

Wilson, R.C., T. Vacek, D.L. Lanier, and D.A. Dewsbury. 1976. Open-field behavior in muroid rodents. *Behavioral Biology* 17:495–506.

Wirtz, R.A. 1980. Occupational allergies to arthropods. *Bulletin of the Entomological Society of America* 26:356–360.

Wolff, R.J. 1985. Mating behaviour and female choice: their relation to social structure in wild caught house mice (*Mus musculus*) housed in a semi-natural environment. *Journal of Zoology*, Series A: Proceedings of the Zooloical Society of London 207:43–51.

Woodmansee, R.B., C.J. Zabel, S.E. Glickman, L.G. Frank, and G. Reppel. 1991. Scent-marking ("pasting") in a colony of immature spotted hyuenas (*Crocuta crocuta*): a developmental study. *Journal of Comparative Psychology* 105:10–14.

Woods, J.E. 1978. Interactions between primary (cage) and secondary (room) enclosures. Pages 65–83 in E. Besch, Chairman, *Laboratory Animal Housing*. National Academy of Sciences, Institute of Laboratory Animal Resources, Washington, D.C.

———. 1980. The animal enclosure—a microenvironment. *Laboratory Animal Science* 30:407–413.

Woolpy, J.H., and B.E. Ginsburg. 1967. Wolf socialization: a study of temperament in a wild social species. *American Zoologist* 7:357–363.

Wrangham, R.W. 1974. Artificial feeding of chimpanzees and baboons in their natural habitat. *Animal Behaviour* 22:83–93.

———. 1980. An ecological model of female-bonded primate groups. *Behaviour* 75:262–300.

Wren, H.N., J.L. Johnson, and D.G. Cochran. 1989. Evolutionary inferences from a comparison of cockroach nuclear DNA and DNA from their fat-body and egg endosymbionts. *Evolution* 43:276–281.

Wright, P.C., D. Haring, M.K. Izard, and E.L. Simons. 1989. Environment for nocturnal primates in captivity—an enlightened approach. Pages 61–74 in E. Segal, ed., *Housing, care, and psychological well-being of captive and laboratory primates*. Noyes Publications, New York.

Wright, P.C., M.K. Izard, and E.L. Simons. 1986a. Reproductive cycles in *Tarsius bancanus*. *American Journal of Primatology* 11:207–215.

Wright, P.C., L.M. Toyama, and E.L. Simons. 1986b. Courtship and copulation in *Tarsius bancanus*. *Folia Primatologica* 46:142–148.

Wyers, E.J., H.V.S. Peeke, and M.J. Herz. 1973. Behavioral habituation in invertebrates. Pages 1–58 in H.V.S. Peeke and M.J. Herz, eds., *Habituation*, volume 1. Academic Press, New York.

Wynne-Edwards, K.E., and R.D. Lisk. 1988. Differences in behavioral responses to a competitive mating situation in two species of dwarf hamster (*Phodopus campbelli* and *P. sungorus*). *Journal of Comparative Psychology* 102:49–55.

Wynne-Edwards, V.C. 1962. *Animal dispersion in relation to social behavior*. Hafner Publishing, New York.

Yaglou, C.P., E.C. Riley, and D.I. Coggins. 1936. Ventilation requirements. *American Society of Heating and Ventilating Engineers Transactions* 42:133–162.

Yerkes, R.M. 1925. *Almost Human*. Jonathan Cape, London.

Yoerg, S.I. 1991. Social feeding reverses learned flavor aversions in spotted hyenas (*Crocuta crocuta*). *Journal of Comparative Psychology* 105:185–189.

Zeuner, F.E. 1963. *A history of domesticated animals*. Hutchinson, London.

Ziegler, T.E., W.E. Bridson, C.T. Snowdon, and S. Eman. 1987a. Urinary gonadotropin and estrogen excretion during the postpartum estrus, conception, and pregnancy in the cotton-top tamarin (*Saguinus oedipus oedipus*). *American Journal of Primatology* 12:127–140.

Ziegler, T.E., A. Savage, G. Scheffler, and C.T. Snowdon. 1987b. The endocrinology of puberty and reproductive functioning in female cotton-top tamrins (*Saguinus oedipus*) under varying social conditions. *Biology of Reproduction* 37:618–627.

Zigman, S., J. Schultz, and T. Yulo. 1973. Possible roles of near light in the cataractous process. *Experimental Eye Research* 15:201–208.

Zuckerman, 5. 1932. *The social life of monkeys and apes*. Routledge & Kegan Paul, London.

Names and Addresses of Contributors to the Volume

Dr. Joseph Annelli
USDA-APHIS-REAC
Federal Center Building, Room 268
6505 Belcrest Road
Hyattsville, Maryland 20782

Dr. Kathryn Bayne
Veterinary Resources Program
National Center For Research Resources
Building 14-D, Room 313
National Institutes of Health
Bethesda, Maryland 20892

Dr. Benjamin B. Beck
Associate Director for Biological Programs
National Zoological Park
Washington, D.C. 20008

Dr. Sue A. Beckley
Departments of Psychology and Zoology
University of Massachusetts
Amherst, Massachusettts 01003

Dr. Emerson Besch
College of Veterinary Medicine
University of Florida
Box J-125, JHMHSC
Gainesville, Florida 32610

Dr. Richard Brenner
USDA-ARS
Medical and Veterinary Entomology Research Laboratory
P.O. Box 14565
Gainesville, Florida 32604

Dr. Gloria S. Caldwell
Department of Psychology
University of California at Berkeley
Berkeley, California 94720

Dr. Maria Inês Castro
Department of Zoology
University of Maryland
College Park, Maryland 20742

Dr. Donald Allen Dewsbury
Psychology Department
University of Florida
Gainesville, Florida 32611

Mr. James Doherty
General Curator
New York Zoological Park
185th Street and Southern Blvd.
Bronx, New York 10460

Dr. Edward F. Gibbons, Jr.[1]
Department of Psychology
State University of New York
    at Stony Brook
Stony Brook New York 11794
    [1]Present Address:
    Center For Science and Technology
    Briarcliffe College
    250 Crossways Park Drive
    Woodbury, New York 11797

Dr. Benson Ginsburg
Department of Biobehavioral Sciences
University of Connecticut
Storrs, Connecticut 06268
            and
Department of Psychiarty
University of Connecticut Health Center
Farmington, Connecticut 06032

Dr. Stephan Glickman
Department of Psychology
University of California at Berkeley
Berkeley, California 94720

Dr. Roy Henrickson
Office of Laboratory Animal Care
University of California at Berkeley
Berkeley, California 94720

Dr. Kay Izard
Research Triangle Institute
P.O. Box 12194
Research Triangle Park, North Carolina 27709

Dr. Charles H. Janson
Department of Ecology and Evolution
State University of New York
    at Stony Brook
Stony Brook, New York 11794

Dr. Andrew King
Department of Psychology
Indiana University
Bloomington, Indiana 47405

Dr. George V. Kollias, Jr.
Department of Clinical Sciences
College of Veterinary Medicine
Cornell Univeristy
Ithaca, New York 14853

Dr. Timothy D. Mandrell
Department of Comparative Medicine
Health Sciences Center
University of Tennessee
956 Court, Box 17
Memphis, Tennessee 38163

Dr. Emil Menzel
Department of Psychology
State University of New York
    at Stony Brook
Stony Brook, New York 11794

Dr. Melinda A. Novak
Psychology Department
Tobin Hall
University of Massachusetts
Amherst, Massachusetts 01003

Dr. K. Daniel O'Leary
Department of Psychology
State University of New York
    at Stony Brook
Stony Brook, New York 11794

Dr. Peggy O'Neill
Animal Center
National Institutes of Health
Dickerson, Maryland 20842

Michael E. Pereira
Duke University Primate Center
3705 Erwin Road
Durham, North Carolina 27705

Dr. Charles Snowdon
Department of Psychology
University of Wisconsin
Madison, Wisconsin 53715

Dr. Michael Stoskopf
Professor and Head
Department of Companion Animal and Special Species Medicine
College of Veterinary Medicine
North Carolina State University
Raleigh, North Carolina 27606

Dr. Stephen J. Suomi
Laboratory of Comparative Ethology
Building 31, Room B2B15
National Institute of Child Health and Development
National Institutes of Health
900 Rockville Pike
Bethesda, Maryland 20892

Dr. Meredith West
Department of Psychology
Indiana University
Bloomington, Indiana 47405

Dr. Everett Wyers
Department of Psychology
State University of New York
    at Stony Brook
Stony Brook, New York 11794

Index

A

Adaptates, definition of, 78

Agricultural Animal Guide, importance of psychological well-being, 47, 48; review of, 47, 48

Agricultural Animals, housing and environmental requirements for, 79

American Association For Accreditation of Laboratory Animal Care, accreditation council, 72

American Museum of Natural History, rats maintained in an outdoor environment, 212

American Psychoanalytic Association, panels at annual meetings, 69

Animal and Plant Health Inspection Service, administration of Animal Welfare Act, 51; legal responsibilities to follow up noncompliance violations, 58

Animal Behavior, abnormal behavior as a criterion of psychological well-being, 219; acquisition of vocal behavior in primates, 225; adaptive importance of dominance hierarchies, 187; affiliative behavior in voles, 192; aggression, 241–243; anthropomorphic assessment, 21; as an index of health, 24, 155; as

indicators of the adequacy of captive environments, 209; behavioral stereotypies, 211; behavioral totipotentiality, 154; brood parasitism in cowbirds, 163, 164; burrow building in rodents, 26, 182, 215; communication, 80, 229; comparative studies of social behavior, 276; concern for animal care and use, 64; copulatory postures of female cowbirds to males' songs, 169–172; effects of crowding on, 186; effects of isolation rearing on social behavior, 68, 69; flight distance, 82, 83; foraging, 134, 135, 183, 226, 240, 243, 244–247, 274, 275, 284; genetic basis of behavior, 191, 192, 275, 276; habits, 154; herding behavior of dolphins, 7, 8; importance of learning in development, 275; influence of kin relationships, 273; influences of photoperiodicity, 89; interaction between genotype and environment, 275, 276; knowledge of ontogenetic mechanisms, 223; locomotor activity, 240, 241; maintenance of homeostasis, 90, 91; mate choice, 194, 195; normative species-typical behavior, 22, 154, 223; opportunities in naturalistic captive environments, 128;

H

reared rhesus monkeys, 253, 254, 257; should demonstrate high reproductive success in animals, 219, 220; space considerations, 81, 83; special importance of epizootic and zoonotic diseases, 92–94; species-typical behavior, 24–26; specific requirements for each primate species, 223, 224; success of reintroduction as an evaluation criteria of, 222; techniques for studying behavior and physiology, 218, 219; testing for auditory perception in, 233; testing for learning capacity and visual perception in, 232, 233; understanding how arthropod populations flourish in, 98, 99; use of naturalistic substrates in indoor rooms for prosimians, 116–119; used to house golden lion tamarins, 262; vegetation to be avoided, 80; veterinary criterion of psychological well-being, 219; vs. laboratory settings, 24; what is a naturalistic research setting? 26

Naturalistic Facilities (*see also* Naturalistic Environments), administrative, regulatory, and technical obstacles, 5; communication of research and scientific needs, 287; natural materials, 284; non-standard features, 1, 2; scientific and management objectives, 285; scientific rationale for non-standard housing, 2–4; standards, 4, 5; what constitutes a, 283, 284

Natural Selection, animals' ability to adapt, survive, and reproduce, 79; homeostatic adaptations, 77, 78, 90

New York Zoological Park, Jungle-World, mangrove swamp exhibit, 134, 135

New York Zoological Society, St. Catherine's Island, ring-tailed lemur psychological cages, 268

NIH Guide for the Care and Use of Laboratory animals, facilities requirements for research, 63, 64; Institute for Laboratory animal Resources (ILAR), 63; major guideline for animal use, 62; Part IV of the Federal Register, 63; recommendations, 63; vertebrates, 63

Nobel Prize, 67

Non-Invasive Research Techniques, in naturalistic environments, 232, 234

Norway Rats, effects of crowding on social behavior, 90, 91, 186; stability of breeding groups in, 185, 186

O

Old-Field Mice, behavior in semi-natural environments, 189, 190

Orangutans, as a solitary species, 249; response to manipulable objects, 248; social aspects of, 111; social housing in zoological parks, 250

P

Peking Ducks, social behavior, 31

Pesticides, use in controlling arthropod pests, 103

Pests, arthropods and vermin, 285; behavior and ecology, 285; "CIA" concept of suppression, 102, 103; cockroaches, 288; concepts in the management of, 96, 97; number of cockroach species, 97; strategy for controlling established infestations,

mals, 24, 25; in which a species evolved, 239; rehabilitation of, 222; selecting habitats to be modeled in the laboratory, 239, 240; selection of features for simulation, 240

Wild Mice, behavior in naturalistic environments, 27

Wisconsin General Testing Apparatus (WGTA), used to study learning sets, 232

Wolves, isolation studies, 69; observations in captivity, 69, 70

X

Xenobiotic Metabolizing System, analysis of xenobiotic compounds, 149; definition of, 149; net metabolizing ability of animals, 149

Z

Zoological Parks, Apenheul Zoological Park, 267; behavioral observations in, 68; behavioral stereotypies of animals in, 211; concern for humane treatment of animals, 64; determination of quality and reputation, 79; inclusion of research in the master plan, 81; maintaining animals in complex habitats, 37; National Zoological Park, 262, 267; New York Zoological Park, 134, 135, 268; out-dated environments, 128; Philadelphia Zoological Park: trauma and mortality in birds and mammals, 93; primate social housing in, 250; research in, 218; responsibility not to import wild-caught primates, 219, 220; use of gunnite, 104; use of naturalistic habitats to house golden lion tamarins, 262; zoo environments as compared to animals' native environments, 79; zoonoses in, 93

Zoonoses and Disease In Animals, associated with animal environments, 92–94; of arthropods, 98; of cockroach, 98; types of diseases found in animals, 93